HOT CARBON

JOHN F. MARRA

HOT CARBON

—

Carbon-14 and

a Revolution

in Science

COLUMBIA UNIVERSITY PRESS
NEW YORK

Columbia University Press
Publishers Since 1893
New York   Chichester, West Sussex
cup.columbia.edu
Copyright © 2019 Columbia University Press

Library of Congress Cataloging-in-Publication Data
Names: Marra, John F., author.
Title: Hot carbon : carbon-14 and a revolution in science / John F. Marra.
Description: New York : Columbia University Press, [2019] |
Includes bibliographical references and index.
Identifiers: LCCN 2018044692 | ISBN 9780231186704 (cloth : alk. paper) |
ISBN 9780231546782 (e-book)
Subjects: LCSH: Carbon—Isotopes. | Radiocarbon dating. | Radioactive dating.
Classification: LCC QD181.C1 M3875 2019 | DDC 551.7/01—dc23
LC record available at https://lccn.loc.gov/2018044692

Columbia University Press books are printed on permanent
and durable acid-free paper.

Printed in the United States of America

Cover design: Noah Arlow

CONTENTS

Preface vii
Acknowledgments xi

PROLOGUE: ABOARD THE RESEARCH VESSEL *ENDEAVOR*,
SOUTH OF ICELAND, MAY 1991 1

1. DISCOVERY 5

2. DISCOVERY'S WAKE 25

3. THE "INVISIBLE PHENOMENON" 38

4. DATING 51

5. PHOTOSYNTHESIS 71

6. CALVIN'S CYCLE 81

7. SCINTILLATIONS AND ACCELERATIONS 102

8. THE SHROUD OF TURIN AND OTHER RELICS 116

9. OCEAN CIRCULATION 133

10. CARBON-14 IN THE OCEAN 150

11. OCEAN FERTILITY 167

12. RESOLUTION: PLANKTON RATE PROCESSES
IN OLIGOTROPHIC OCEANS 183

13. CARBON-14 AND CLIMATE 201

EPILOGUE 223

Appendix 1. List of Nobel Prize Winners Mentioned 227
Appendix 2. The Periodic Table of Elements 229
Notes 231
References 243
Index 249

PREFACE

"**S**eldom has a single discovery in chemistry had such an impact on the thinking in so many fields of human endeavour. Seldom has a single discovery generated such wide public interest." So declared the chair of the Nobel Committee, quoting one of the nominating scientists, in awarding the 1960 Nobel Prize in Chemistry to Willard Libby for his method of age determination using carbon-14. A year later, Melvin Calvin received the Nobel Prize in Chemistry for unraveling the path of carbon fixed in photosynthesis. The presentation speech for Calvin's Nobel noted that "photosynthesis is the absolute prerequisite for all life on earth and the most fundamental of all biochemical reactions," adding that "the radioactive carbon isotope, ^{14}C, well-known also in other connections, has played a particularly important role."[1]

The stories surrounding the discovery and applications of carbon-14 in the second half of the twentieth century are the subject of this book. R. J. Gordon, in his *Rise and Fall of American Growth: The U.S. Standard of Living Since the Civil War*, notes that, except in the rural South, the daily life of every American changed beyond recognition between the end of the Civil War and the end of the Great Depression. By the end of World War II and the beginning of the so-called atomic age, the daily life of all Americans had changed profoundly. Carbon-14 is part of that sea change in outlook and life, and one of the most significant products of the atomic age.

Despite being so worthy of attention, however, the role of carbon-14 became submerged, except in scientific circles. Biological metabolic cycles, displayed on wall charts, never reference the isotope that allowed those transformations to be portrayed. The popular press refers to "carbon dating" or "carbon dates" without noting the long-lived radioisotope that makes fixing those dates possible. Much research in oceanography involves carbon-14, but this is not among the topics that inspire colorful documentaries on public television. I hope with this book to uncover some of the secrets involved in the history of carbon-14.

Stories of the impact of carbon-14 on daily life are told here quasi-chronologically, culminating in a continuing relationship with humanity and its advancement. I begin with the discovery of carbon-14 at the Berkeley Radiation Laboratory (chapters 1 and 2), then consider the discovery of radioactivity itself and how it is measured (chapter 3). Not long after the discovery of carbon-14 in the lab, it was found to be produced in the upper atmosphere, naturally and continuously. Chapter 4 describes how the natural occurrence of carbon-14 allows the dating of anything that accumulates carbon during its lifetime, tracing Willard Libby's early efforts to establish the method. Chapter 5 recounts the history of the study of photosynthesis, and chapter 6 tells how Melvin Calvin and his team, in the late 1940s, worked out how carbon dioxide, via photosynthesis, is assimilated into organic matter. Chapter 7 returns to the assessment of carbon-14 (covered initially in chapter 3)—in particular, the inception of accelerator mass spectrometry (AMS), without which radiometric dating using carbon-14 would be next to useless. Chapter 8 then discusses various applications of carbon-14, using AMS, in dating archeological and other historical relics, the peopling of North America, and the age of human remains at archeological sites.

In chapters 9 and 10, I consider circulation in the deep ocean as it was worked out in the 1980s, the Earth's carbon cycle, and how the carbon cycle affects Earth's climate. Chapters 11 and 12 review the problem of ocean productivity. In chapter 11, I jump back to the late 1940s and early 1950s to reappraise the introduction of carbon-14 to biological oceanography. I cover the early debates (still continuing today) about the ocean's productivity and what led to the project "Plankton Rate Processes in Oligotrophic Oceans," the topic of chapter 12. That project, in the mid-1980s, resulted in a paradigm shift in the study of ocean productivity—its magnitude and its importance to the Earth's carbon cycle and ocean resources. I saved the study of carbon-14 and climate for chapter 13. Deciphering climate conditions since the last deglaciation, which is about the limit for the application of carbon-14, began in earnest in the late 1980s and spilled over to the twenty-first century. New data and ideas are rife at present, being generated almost monthly.

As long as carbon constitutes the stuff of life on planet Earth, carbon-14 will find application. Carbon is what we're made of. Carbon is life. It is fundamental also to how we live, how we interact with the Earth's spheres, how the Earth is habitable—pretty much everything. And since the discovery of a long-lived radioisotope of carbon, we have an amazing tool to delve into almost every aspect of existence on Earth—and perhaps the universe. Sir James Jeans, an early-twentieth-century physicist and astronomer, once said, "Life exists in the universe only because the carbon atom possesses certain exceptional qualities."

What we'll find in this book is that carbon-14 is ubiquitous; it can be measured everywhere life occurs, or has left an imprint. Ubiquity is both a good and a bad thing, scientifically. Teasing apart the variables and drivers, we find that, in most cases, the uses of carbon-14 with the clearest results are those in which carbon-14 is added to a system in comparison with distributions that occur naturally. An exception is the use of carbon-14 in dating archeological artifacts, although here, too, corrections have to be applied, and these become a science in themselves.

For myself, I was part of perhaps the last generation of oceanographers who worked entirely as "soft money" scientists. "Hard money" meant being in a salaried position at a university or government lab. As soft-money scientists, we depended for our livelihood entirely on government grants or contracts. As one of my lab directors once said, we "lived by our wits." We acted like small businesses, proposing our ideas to funding agencies ("selling" seems too strong a word), and indirectly to our peers. If successful, we implemented budgets, hired personnel (technicians, administrators), perhaps enlisted students and postdoctoral fellows. Eventually, in completing the research, we produced the product—a scientific contribution, a scholarly work that advanced knowledge of (in my case) the ocean. The Lamont-Doherty Earth Observatory, where I spent most of my career, had about 75 to 100 soft-money researchers at any one time; unlike in many academic departments, these researchers were young, with an average age of 35. Soft-money science was the model for oceanographic research in the second half of the twentieth century. Academic programs were few, while government needs for understanding the ocean were many. One of the admirals in charge of the Office of Naval Research once said, "If it happens in the ocean, we want to know about it."

ACKNOWLEDGMENTS

First, I thank J. Cherrier, N. McCabe, and J. Wyatt for giving me important feedback on various chapters. It was a great help. I am also grateful for comments, corrections, and suggestions on chapter 13 from D. Peteet and D. V. Kent. Any remaining errors are my own.

The stories about carbon-14 in which I was personally involved all occurred in the early to mid-1980s, not long after I finished graduate school. During that time, several individuals helped start me on my research program. Tom Malone, now an emeritus professor at the University of Maryland, convinced the director of the Lamont-Doherty Earth Observatory that it would be a good idea to expand his small group of biological oceanographers amid a sea (!) of geophysicists, seismologists, and geochemists, the overwhelming majority of the research staff at the observatory. The September day in 1977 that I moved into my office at Lamont as a postdoctoral fellow, another junior scientist, Jim Bishop, came and asked me if I wanted to be part of a proposal. Naturally, I said yes, without even knowing the details. Pierre Biscaye, a geochemist and a member the observatory's senior staff, became our champion for that project, persuading the program managers at the Office of Naval Research (on a site visit to Lamont) to take a chance on three young scientists (myself, Jim, and Wilf Gardner). A few years later, Dick Eppley, perhaps the best-known biological oceanographer at the time, invited me to help him with a project, "Plankton Rate Processes in Oligotrophic Oceans," the topic of chapter 12, whose objective was to resolve the controversies surrounding the use of carbon-14 in biological oceanography. Somehow I became a member of his advisory committee for the project, and we were known as "The Three Johns": John McNeill Sieburth, John Heinbokel, and me.

In 1982, Pete Jumars and Mary Jane Perry, from the University of Washington, began two-year rotations as program managers in Washington, D.C., at the Office of Naval Research (ONR) and the oceanography program at the National Science Foundation (NSF), respectively. In my opinion, Pete and Mary Jane (along with Bob Wall at NSF) transformed, or at least reclaimed and revitalized, biological oceanography at the NSF and enhanced its presence for the Navy. One project that Mary Jane recommended for funding was the project in which Dick Eppley invited me to participate: "Plankton Rate Processes in Oligotrophic Oceans." Eric Hartwig, who took over from Pete after his rotation at ONR ended, oversaw the first five-year "Accelerated Research Initiative" (ARI) awarded to biological oceanography, a collaboration between ocean optics and ocean biology. (Pete was the one who had earlier proposed the idea to the admiral.) Over the next several years, Eric carefully shepherded that first ARI through the not-always-cooperative vested interests—but that's a story for another book. Early in my career, these colleagues and programs helped stabilize my research program at Lamont. In my personal life during those years, I enjoyed the support of my wife Glenna and our two children, Benjamin and Natasha. By being there, you guys got me through those oftentimes stressful years.

Finally, I note the passing of Wally Broecker, who figures prominently in chapters 9, 10, and 13. Wally not only pioneered research with carbon-14, he spearheaded ocean and climate science over the past 60 years.

HOT CARBON

PROLOGUE

W e looked out through the *Endeavor*'s windows from the safety of the ship's main lab, apprehension showing on our faces, knowing we had to go out on deck. The recovery team consisted of my colleague, Chris Langdon; my lab tech, Carol Knudson; Juan Soriano, a tech on loan from an instrument company located in Miami; and me. The seas bordered on chaotic. The wind had kicked up over the day to 55 knots, and the seas responded according to merciless physics. For Chris, Carol, and me, this was another day on the job. Juan, however, was on his first ever oceanographic cruise, and the look on his face showed that a blustery and foul sea was not what he had signed up for. Still, he was game. And regardless, we had a spar buoy harboring plankton samples and irreproducible data to retrieve from those seas. We were already invested in all the work done the previous evening to prepare the experiment. There was no question about recovering the buoy.

The *Endeavor* occupied a station to the south of Iceland, an unforgiving part of the Atlantic in the best of seasons. We, the scientists, were researching relationships among the plankton types in response to the ocean's daily and weekly whims. My group's particular job was to measure how much photosynthesis was happening in the resident phytoplankton populations in the ocean's surface layers.

Eighteen hours before, we had collected water samples from various ocean depths, filled special flasks with the water from those depths, spiked them with the radioactive isotope carbon-14, and attached the flasks to a long line according to the depths at which they had been captured. We attached the top of the line to a free-floating 15-foot spar buoy that had slack lines along its length that could snag a grappling hook, aiding retrieval. At the top of the buoy we fixed a flashing light. Having everything set, we eased the weighted line over the side, finally casting the buoy overboard to float by itself away from the ship. That was at dawn—at that time of year, about 4:30 AM. The idea was to let the phytoplankton in the samples photosynthesize during the long spring day, taking up the inorganic carbon-14 in the process. Now, just after sunset, we needed the samples back to see how the phytoplankton had done.

Endeavor is a midsize (165-foot) research vessel, operated by the University of Rhode Island, and exceptionally seaworthy. Captains of oceanographic ships can handle 55-knot winds, and even much more severe weather. *Endeavor*'s captain, Tom Tyler, had logged thousands of sea miles. By the time we planned the recovery, it was dark, or as dark as it gets in the month of May at 60 degrees north latitude. We zipped up our work vests and stepped out on deck, balancing with the rolling of the ship. The work vests had long straps that we could clip to a railing along the ship's outer bulkhead, tethering us to the ship. If we slipped on the deck, we might get bruised, but we weren't going overboard.

We were ready in our persons, and then gaped, open-mouthed, at the seas. A wall of water looked to crash into the ship, and us, broadside. Captain Tyler maneuvered so the wave went under the ship instead of over the railing, but that meant a heavy roll. In the trough of the wave, the captain had our spar, with its light flasher, in sight ahead off the ship's starboard bow. But we had a problem. The grappling hook and line that we needed to snag the buoy were hopelessly tangled. We were frantically trying to untangle the line, while over our shoulders we saw our buoy just ahead of the midpoint of *Endeavor*'s starboard side and moving aft. Then we heard screaming from the bridge wing. The captain was ready to retrieve the buoy. We scientists were not. Captain Tyler belted out the most colorful language about our abilities and ancestry, to the effect of "Get that damned hook ready, you . . ." He had a point. He had impressively positioned the ship just right in those 10-foot seas so we could recover the buoy—and we weren't ready. He'd done his job; we'd failed at ours.

Amazingly, he was able to maintain the ship's position relative to the buoy for a few more precious seconds while we untangled the line, still experiencing the captain's justifiable fury. Then the toss, and luckily the hook grabbed that first time, and held. As the ship rolled toward the buoy and then back away, we could use the roll and the seas to leverage the buoy up to the rail and muscle it aboard, all the while trying to keep our balance against the ship's lurching with the waves. We pulled aboard the rest of the line with the attached bottles and brought the whole affair inside a bulkhead and out of the weather. Huge sighs of relief. The intrepid oceanographers had another data point in hand—a completed experiment, or at least the hard part.

Later that night, the winds increased to 74 knots, technically qualifying this storm as a hurricane, if only briefly. When the readout of the meteorological data on the ship's computer recorded the wind hitting 74 knots, someone hit the "print screen" button, and we all got a souvenir from the cruise—a badge of honor.

While the weather worsened, Carol and I got to work, funneling the seawater from the bottles through small glass-fiber disks, thereby separating out and collecting the particulate matter containing the photosynthesizing phytoplankton. We stored the disks in vials, later to assay their radioactivity back at the lab.

For the next 18 hours, the ship was "hove-to" to ride out the storm. No work could be done on deck, no gear put over the side and into the sea. Winds of 55 knots with 10- to 15-foot seas were the worst—or, as I thought at the time, the most exciting—conditions for doing a measurement of the daily photosynthetic production in the ocean. This is not to say that experiments like this never go wrong. Each of my failures in data collection is burned into my memory as a lost opportunity for understanding. In oceanography done from ships, there are no second chances. In the "bench sciences," if an experiment doesn't work, you can come back to the lab the next day for a redo. That can't be done at sea. Each event, each experience, each observation, is unique. Being out on deck that night, exposing ourselves to the storm-tossed seas, was worth the risk.

As I discuss in a couple of the later chapters of this book, carbon-14 allowed us to make those measurements from the *Endeavor* that would otherwise be impossible. Measuring the photosynthetic assimilation of carbon in the ocean, the first step in the ocean's food web, occupies a small slice of where carbon-14 has been critical to the advance of science. The story of carbon-14 begins about 50 years before our *Endeavor* cruise, and the beginnings of carbon-14's application in oceanography about 10 years after that, in 1952.

The other applications of carbon-14 are as varied as they are groundbreaking and far-reaching. The discovery of carbon-14 in 1940, its application to the dating of artifacts, the working out of carbon dioxide assimilation into organic matter via photosynthesis, giving time scales in oceanography and chronologies to climate—all involve dramatic change in understanding and perspective: revolutions in twentieth-century science.

1

DISCOVERY

The police officers in the patrol car eyed the figure hunched over in the rain, his gait a little wobbly, and pulled up beside him. He certainly looked suspicious. Rumpled, red-eyed, and with a three-day growth of beard, he was, in a word, a mess. The Berkeley police were on the lookout for an escaped convict who had committed a multiple murder the previous evening; maybe now they had their culprit. They shoved him into the patrol car—he was in no condition to resist—and drove to the station. The suspect was put in front of a survivor of the crime. The two individuals—one distraught, the other tired and disheveled—stared blankly at each other. Clearly, neither knew who he was looking at. After some further questioning, the police let their suspect go. In the early morning hours of February 27, 1940, Martin Kamen could finally go home to bed.

Kamen had been working three nights straight inside the Berkeley Radiation Laboratory, bombarding graphite targets with deuterons, atomic particles made up of a proton and a neutron, produced by the Berkeley cyclotron. Kamen's work was relegated to nights because of the daytime needs for cyclotron time for producing established isotopes, phosphorus-32 and iron-59, used in cancer therapies. His research was more of a fishing expedition, important but lower priority. He had been at it for more than a month, pasting back pieces of irradiated graphite that had been blasted off the target during the previous night's bombardment and, in so doing, exposing himself to even more radiation. Kamen, frustrated, discouraged, and not a little desperate, stayed up those three nights for a final push.

The third night, torrential rain pelted the windows from a violent storm that would flood low-lying areas of Berkeley and Oakland. There were lightning strikes outside. High-energy cannonades burst from the cyclotron. Screams and moans were broadcast from the recording of a Gallic tragedy played by a French drama class that had convened that night on the mezzanine above the cyclotron control desk. The scene was set. The beam of deuterons from the cyclotron targeted a piece of graphite with the hope of slowly revealing a new form, or "isotope," of carbon. It was not exactly the "It's alive!" moment when Boris Karloff, as Frankenstein's creature, raised his arm, but maybe just as exciting, if delayed. Later that night, Kamen collected the pieces of blasted-off graphite, which looked like black gravel, into a small bottle and left the sample on Sam Ruben's desk.

Finally making it home after his run-in with the police, Kamen slept until midafternoon. That gloomy night and morning of February 27, 1940, began a revolution in physiology, biochemistry, archeology, geology, biomedicine, oceanography, paleoclimatology, and anthropology, as well as nuclear chemistry. Carbon-14, perhaps the most important isotope to life on Earth, was "born."[1]

■ ■ ■

"Isotope" comes from the Greek, meaning "same" (isos) "place" (topos). Isotopes of an element occupy the same square on the periodic table of elements, but have an extra proton or neutron or two, or may be missing one, giving them a slightly heavier or lighter atomic mass. The periodic table of the elements is perhaps the singular achievement of nineteenth-century chemistry, ordering the elements by their atomic characteristics and chemical behavior. Each element is made up of atoms; the atoms are made up of protons and neutrons in a nucleus, surrounded by electrons. The number of protons determines each element's atomic number, and the protons plus neutrons give the element its atomic mass. The elements, so carefully displayed, one to each square on the periodic table, are, in fact, families, with different isotopes of an element crowding into the square. Chemically, they are nearly indistinguishable. But identifying isotopes of an element has allowed tremendous advances in the basic sciences.

Some 98.99 percent of all the carbon in the universe has an atomic weight slightly in excess of 12. The carbon atom has six protons and six neutrons in its nucleus. Orbiting electrons account for the slight excess weight. This is the most common isotope of carbon, carbon-12, or in the symbology of nuclear chemistry, ^{12}C. An isotope with an extra neutron, designated ^{13}C, makes up most of the rest of the carbon in the universe, about 1 percent of the total. Carbon-13 is a stable isotope; it has been around on Earth as long as carbon-12 has, something like 4.6 billion years. The other isotope, carbon-14, much rarer than its siblings, occurs once in a trillion carbon atoms and is radioactive. It decays over time, thereby changing to become carbon's next-door neighbor in the periodic table, nitrogen, with a mass number of 14—seven protons and an equal number of neutrons. As you might guess, ^{14}N, or nitrogen-14, plays an important role in the carbon-14 story.

Radioactivity was discovered by accident in the 1880s, by Henri Becquerel, as energy emanating from a special type of mineral that gives off its own light, a glow called fluorescence (see chapter 3). Later, Marie Curie, a contemporary of Becquerel, named this energy "radioactivity." We now know that radioactivity occurs naturally, especially for elements at the high-numbered end of the periodic table and particularly those heavier than bismuth, with an atomic number of 83. Radioactivity is energy; it is the energetic decay of elements that changes them to other, more stable forms. And like the parents and siblings of human families, each isotope's radioactivity behaves differently—for example, how fast it decays, what part of the atom splits off, and how much energy the isotope liberates in the process. For our purposes, the most important of these behaviors is how fast the isotope decays. In the early 1900s, Ernest Rutherford, then at McGill University in Montreal, came up with the idea of the "half-life"— the time it takes for half the quantity of an isotope to disappear through radioactive decay.

Many people have a bank account of some kind, often an interest-bearing account. The interest is the return banks give you for the use of your money. Over time, without doing anything, your funds increase according to that interest rate. The half-life works more like a spend-thrift who withdraws 50 percent of his funds at regular intervals. If your account has a balance of $1,000, withdrawing half at the end of the first month leaves $500. After another month, withdrawing 50 percent of $500 leaves $250, and so on. For those with a more mathematical mind, a half-life means expressing values in logarithms to the base 2, the logarithm to the base 2 being the exponent of 2. The more familiar logarithms, especially to those of us who remember slide rules, are base 10: $10^1 = 10$, $10^2 = 100$, and $\log(100) = 2$. Logarithms to the base 2, or "log2," allow one to compute things that double, quadruple, or conversely, divide in half, like the half-life.

For example, a doubling of something, like bacterial cells means $\log_2(2) = 1$, or one doubling. When we divide something in half, or $\log_2(0.5) = -1$, that is one half-life. Halving again, or $\log_2(0.25) = -2$, is two half-lives. It turns out that half-lives can range from seconds to millions or even billions of years, depending on the isotope. Given the age of the universe, some of the isotopes originally created after the Big Bang (some 13 billion years ago) have simply decayed out of existence. Another isotope of carbon, carbon-11 illustrates this point. Carbon-11 has five protons and six neutrons, a very unstable mix. Carbon-11 is not "comfortable in its skin" and has a half-life of 22 minutes. It lives for only a few hours before decaying away to boron, carbon's other neighbor on the periodic table.

■ ■ ■

Martin Kamen (figure 1.1) was born in Toronto in 1913. His parents were Lithuanian and Byelorussian émigrés. When Kamen was still a young boy, they moved to Chicago to be near relatives. His father set up a photography business, his mother went into real estate, and soon they were leading a comfortable life in 1920s Chicago. Kamen had a remarkable intellect, skipping grades and finishing high school early. He was a child prodigy on the violin, but switched to the viola in his teens. With the Wall Street crash in 1929, his parents lost much of their middle-class wealth, and they urged him to forego music for something more practical. Two blocks from his home was the University of Chicago, and he attended, majoring in chemistry. His music turned to jazz, and he found he could earn extra money playing in Chicago's speakeasies. After completing his bachelor's degree, he stayed at the University of Chicago for graduate school and in 1936 received his Ph.D. in chemistry, on proton-neutron interactions. He was 27 years old.

FIGURE 1.1

Martin Kamen at the 60-inch cyclotron, September 10, 1939.

Photograph by Donald Cooksey.

With the understanding of radioactivity, isotopes, and atomic structure, nuclear chemistry was quickly taking over from classical chemistry. Elements that had been fixed on the periodic table could now change, and some elements could be created in the lab—notions that would have been heretical only years before. Kamen's dissertation research upended theories about the nature of atomic particles, but the low energies in the accelerators available to him meant that his results would be difficult to verify for another 10 years. Still, becoming proficient in the use of the cloud chamber helped his later research.

The cloud chamber, invented in 1911 by Charles Wilson, had been the prominent mode of investigation into atomic particles from the 1920s to the 1950s. The sealed chamber's internal atmosphere is saturated water vapor. Atomic particles are shot into the chamber and ionize the water vapor, forming condensation nuclei, the "cloud" that reveals the track of the atomic particle as a fine mist. The characteristics of the track are used to identify the particle. An alpha particle—a nucleus of a helium atom stripped of its electrons—makes a broad, straight track. An electron, or beta particle, makes a thin one, and is perturbed by other collisions. Applying an electric field can send differently charged particles in different directions. Later, at Berkeley, Kamen identified tracks in a cloud chamber filled with water vapor and nitrogen gas as resulting from carbon-14.

Although he opted for the cloud chamber, Kamen also pursued another kind: chamber music. Science became his profession, but he carried his viola everywhere. Playing chamber music with other musicians and friends added up to his perfect evening. Later he counted the famous violinist Isaac Stern as a close friend. Kamen was gregarious, very open, and in love with life—a socializer and an extrovert.

In 1935, he endured the tragedy of losing his mother in a car accident. Also, although successful there, Kamen never adapted to the rigid academic environment at Chicago. So, between personal misfortune and professional strains, Kamen looked to move elsewhere. Right after getting his Ph.D. in 1936, and on the advice of his thesis adviser, Kamen left Chicago to visit E. O. Lawrence's Berkeley Radiation Laboratory at the University of California. Lawrence (called E.O.L. by his staff), the lab's director, was gathering together nuclear physicists to work with his new and powerful atomic particle source, the cyclotron. Hoping for a position there, Kamen arrived in the Bay Area with enough money to last him for about six months. Even with precarious finances, compared to the weather in Chicago, Kamen felt as though he had entered heaven. He used his time well, making friends, playing music, and meeting Esther Hudson, who would become his wife. Kamen's sociability, easy camaraderie, interest in politics, relish for science, and outspokenness worked against him later on. Late into his six-month respite, Kamen wrangled what today we would call an unpaid intern position at the Berkeley Radiation Laboratory. Figuring out the cause of an error in an experimental result got him onto the payroll.

Lawrence (figure 1.2) and one of his associates had done experiments on platinum bombarded with deuterons from the cyclotron. They put stacked foils of platinum in the path of the deuteron beam and measured the loss of energy through the stack. Instead of a smooth, simple decline passing through each platinum foil, there were discontinuities, or bumps, in the energy curve. Lawrence considered the discontinuous energy loss to be a

FIGURE 1.2

E. O. Lawrence at his desk.

Photograph from the Lawrence Berkeley Lab Nobelists, www.lbl.gov/nobelists/1939-ernest-orlando
-lawrence/.

major result of cyclotron research. J. Robert Oppenheimer was enlisted to
devise a theory to explain the bumps, and he did so, making a presentation to a packed audience that included a visiting Niels Bohr, the Danish
physicist and 1922 Nobel Prize winner who was responsible for the current
model of the atom. Bohr questioned the results, which contradicted some
of his own unpublished experiments. He pointed out that Oppenheimer's
theory, while elegant, might be useless.

To have their work criticized by the famous Niels Bohr shocked Lawrence, especially as it came after another recent experience with erroneous
experimental outcomes. He was in line for a Nobel himself for developing
the cyclotron, and another misstep might cost him the prize. Lawrence
desperately wanted to clear the matter up, and he called on one of his
faculty members, Ed McMillan, to create a team to investigate the result.

Kamen was chosen for his radiochemical background, and McMillan, presciently believing chemistry important, also chose a young graduate student from Berkeley's chemistry department, Sam Ruben.

Samuel Ruben (figure 1.3) received a Ph.D. from the University of California in chemistry, the same field as Kamen, but emphasizing biological processes. Andrew Benson, whom we will meet in chapter 6, described Ruben as gentle and a quintessential experimentalist. Kamen credits Ruben with "almost single-handedly" spurring interest in the use of isotopes of elements to trace chemical and biological processes during the years leading up to the discovery of carbon-14. He could take an idea—whether or not the idea was originally his, and whether or not he was familiar with the subject—and quickly design the decisive experimental test.

FIGURE 1.3

Sam Ruben.

Photograph courtesy of the Seaborg Archive, Lawrence Berkeley National Laboratory Image Library.

To validate or invalidate Lawrence's bumpy curve of energy loss, the team—McMillan, Kamen, and Ruben—repeated the earlier experiments, and tested the pretreatments of the platinum foil for removal of trace contaminants. Solving the problem meant exhaustive work—exhaustive physically because of the long hours in the lab, and exhaustive scientifically to ensure that all possibilities were covered. Their labors were rewarded with a solution, but not one Lawrence might have liked. Despite all precautions, one contaminant had not been removed during a purification step for the platinum, but had actually been added in: laboratory dust. The dust had been baked into the foils when the foils were flamed. The bumpy energy curve was an artifact.

The effort to test Lawrence's earlier experimental results had three important outcomes. The first was scientific: the discovery of widespread nuclear isomerism—the creation of different excited states of atomic nuclei with various but distinct lifetimes. The second and third outcomes had to do with the working relationships at the Radiation Lab.

As a result of the imbroglio with Bohr, there was now an appreciation of radiochemistry, and chemistry itself, in contributing to the science of the Radiation Laboratory. Chemistry and biology, as scientific disciplines, had always been considered handmaids to nuclear physics. Physics was the pure science where all the secrets of nature were to be divulged, and nuclear physicists were the high priests. The physicists at Berkeley habitually asserted their right to first authorship on research papers. The physicists invariably took precedence for use of the cyclotron. But now the physicists had to concede that chemistry, at least, was important.

The demand for radioactive isotopes also grew, and Kamen became the resident chemist handling calls for radioisotopes from around the country. Though World War II raged on, he was charged with meeting requests for isotopes from Europe as well. These demands, coupled with a lack of interest in radioisotopes within the biology department at Berkeley, made Kamen even more indispensable to the Radiation Laboratory. As a bonus, for correcting his earlier experiments and saving his reputation, Lawrence offered Kamen a staff position as a research fellow, at $1,200 a month. Gratified and delighted, Kamen could now establish himself in the Bay Area, not only scientifically, but culturally and socially as well.

The third outcome from the platinum foil problem was that Kamen and Ruben became acquainted and established a close collaboration and friendship. At the Radiation Lab, Kamen oversaw the production of isotopes from the cyclotron, and Ruben handled the application of those isotopes in chemical research.

There was another reason for this division of labor. After all his work with radioactive isotopes, Kamen found himself a hazard to the experiments. During one observation, a variable background radiation occurred that seemed to be correlated with Kamen's movements around and away from the assay apparatus. Kamen removed articles of clothing without changing the background values. It was finally pinpointed to the front of his pants. Not as bad as you might think—the maximum radioactivity was coming from the zipper in his fly. His body, though, was also above background, and particularly after that episode, I'm sure he heard "Martin, please stand over there" on many occasions. In his partnership with Ruben, given his threat to experimental results, Kamen confined his work to isotope production, with Ruben handling the chemical assays. Their partnership was further divided on the Berkeley campus. Ruben worked in the "Rat House," an adjunct of the chemistry department named for the previous (and perhaps some current) occupants, used for studies by the biology department. Kamen stayed quarantined a few hundred feet away in the Radiation Laboratory along with Ed McMillan, who continued his work on improvements to the cyclotron.

The cyclotron, developed by Lawrence, was the centerpiece of the Berkeley Radiation Laboratory. The idea for a cyclical particle accelerator originated in Germany in the 1920s, but it was not until 1929 that Leo Szilard applied for the first patent. Szilard, a Hungarian, originated many scientific instruments and ideas, including the electron microscope, the chemostat (a system for continuous culture of microorganisms), and the concept of feedback inhibition in metabolism. However, it was Lawrence who fabricated the first working cyclotron in 1934.

A cyclotron accelerates particles to a few percent of the speed of light—only a few percent, but that's still pretty fast. If you want to smash nuclei of atoms to see what happens, or create new nuclei or ions, these particle speeds, and therefore energies, become necessary. At the time, "tron" suffixes—based on the vacuum tube's use of elec*trons* to amplify signals—were added to anything smacking of new technology, mainly as a marketing stratagem. The business-savvy Lawrence always angled his laboratory toward marketing its new products, so the "tron" suffix a natural choice.

The cyclotron (figure 1.4) accelerates atomic particles using two giant hollow electrodes, each shaped like the letter D, called "dees." The dees face each other along their straight sides. The curved part of the facing "letters" then give the device a circular shape. High voltages are applied as alternating current at radio frequencies. The particles themselves are generated by an ion source, a familiar example of which is the cathode ray tube, or old-style TV. The ion source emits particles to the cyclotron at the center point between the dees. The particles are then whipped into a circular trajectory by the magnetic field produced by the electromagnets. The dees alternately become positive and negative as the particles cross the

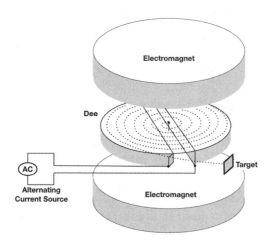

FIGURE 1.4

Schematic of a cyclotron showing the two electromagnets sandwiching the two "dees," an alternating current (AC) source, and the target. Subatomic particles are accelerated around the dees when they are subject to the current, alternating at radio frequencies, and the powerful electromagnets keep the particles entrained to a spiraling outward trajectory (dashed line) as they are accelerated toward the target. Modified from *Symmetry* magazine.

gap between them, increasing the particles' speed. As they speed up, the particles circle and travel outward, forming a spiral trajectory, going through several accelerations along the way. When they reach the outer limits of the dees, they exit and are directed toward a target. Two electromagnets, situated on the top and bottom of the dees, create a magnetic field to keep the particles confined while they are accelerated. The acceleration of the particles depends on the magnetic strength and the radius of the dees: the larger the dees, the greater the energy and acceleration.

The unit of measurement in the cyclotron, and in nuclear physics in general, is the electron volt (eV), which is the energy gained by an electron when it is accelerated through an electric potential difference of one volt. Typically, the electron energies generated by the early cyclotrons were in the millions of electron volts, or MeV. The other unit is the microampere hour, which expresses the number of atomic particles (the current), in amperes (amps), fed into the cyclotron. So, the two units of measure express the number of particles and the particles' energy.

Lawrence's first machine, initially used by Kamen, had a diameter of 27 inches. A year later, Lawrence and his helpers built a 37-inch machine, and, in 1938, a 60-inch-diameter cyclotron. The cyclotron later used by Kamen to discover carbon-14 was the 37-incher, slightly less than a meter across, with magnets that weighed more than two tons. Cyclotrons were remarkable machines for their time, and Lawrence received a Nobel Prize in 1940 for his development of them. The awarding of Lawrence's Nobel is part of this story.

As amazing as these machines were, technicians and their Ph.D. supervisors spent long hours keeping them running at peak performance. I remember a colleague who earned his Ph.D. as a high-energy physicist in the late 1960s saying that for all his graduate education in physics, he was chiefly a plumber and an electrician. Most of the work in high-energy physics means being an expert at these more quotidian skills. (My colleague, soon after a postdoctoral fellowship in nuclear physics, switched fields to oceanography.) Accelerators need continual high-vacuum environments, with a large electrical supply. In an early demonstration of today's connectedness, Lawrence kept a radio at his bedside tuned to the cyclotron's frequency of operation so he could keep tabs remotely. It came to be that he could not sleep without the hum; interruptions would wake him up, warning of a breakdown. His wife must have thought he was more married to the machine than to her.

In 1937, Emilio Segré, director of the physics laboratory at the University of Palermo, Sicily, paid a short visit to Lawrence in Berkeley and took home a souvenir: a strip from the molybdenum shielding used on the cyclotron. After careful measurements and theoretical analysis, he and a colleague showed that certain anomalous radiation coming from the shield material originated from a previously unidentified radioactive element, which they called technetium, filling one of the empty squares (number 43) in the periodic table. More important, technetium is extremely rare on Earth. Its half-life indicated that any primordial technetium would have decayed long ago.[2]

Segré's discovery meant that the cyclotron could realize the alchemist's dream. Historically, two aims of alchemical research were to change one element into another (transmutation) and to overcome the ancient scourge of disease. Pretty much everyone learns in high school chemistry that alchemy was doomed to failure. But now, particle accelerators like the cyclotron might be used for both purposes. Segré's discovery precipitated a change in outlook at the Berkeley Radiation Lab. The cyclotron could be a modern analog to the philosopher's stone, and in producing radioisotopes, it could also cure disease. Materials would be bombarded by accelerated particles, thereby changing one element (or isotope) into another. The transmuted isotopes could be used as therapeutic agents or allow further studies in biology and biomedicine.

Ironically, the alchemist approach to nuclear physics drove the funding for the Berkeley cyclotron. Lawrence expanded the use of radioactivity and isotope production by making bigger, more powerful cyclotrons. The expansion meant higher operating costs and more time spent searching for funds, obtained by promoting and advertising the use of isotopes in medical diagnostics and pharmaceuticals. Today, an emphasis on biomedicine to support atomic energy research seems bizarre for what was to become a crucial laboratory in high-energy physics and nuclear chemistry. But the creation of the National Science Foundation was a decade or more off into the future, as was the Atomic Energy Commission (later to become the U.S. Department of Energy).[3] Lawrence had to find research funds where he could, and he believed his best bet to be the medical industry and foundations predicated on curing human disease.

The cyclotron was 10,000 to 100,000 times more powerful than the machines Kamen was used to working with in Chicago. There, he needed years of continual work to be able to get data on a few hundred particle tracks in the cloud chamber. Could the cyclotron be used to find carbon-14? By the end of 1937, carbon-14 had been identified in cloud chamber experiments. That observation helped define the way the cyclotron would be used, but nothing was known of carbon-14's physical or chemical characteristics.

The decay in the cyclotron of nitrogen was considered first, as nitrogen-14 is the most abundant isotope of nitrogen and nitrogen (as N_2) is the most abundant gas in Earth's atmosphere. Kamen figured that bombarding nitrogen-14 with neutrons in the cyclotron could result in a couple of isotopic outcomes. It might produce a neutron from the nitrogen nucleus, splitting into beryllium-10 plus helium-4, which adds up to 14. In another scenario, neutrons could be added to the nuclei of carbon-12 or carbon-13, eventually creating carbon-14. Given the energies of these reactions, Kamen and his cohorts concluded that the most likely result would be a neutron emission from nitrogen-14 and the formation of carbon-14.

Kamen and his colleagues also figured—or better, guessed—that carbon-14 would not only be unstable—that is, radioactive—but would have a very short half-life. They based these suppositions by analogy on other known isotope pairs having the same atomic number—that is, isobaric—and in the same region of the periodic table. For example, of the isobaric pair boron-10/beryllium-10, beryllium-10 is unstable but has an exceptionally long half-life at 1.4 million years. Another isotope pair, helium-6/lithium-6 has one isotope that is unstable, helium-6, with a half-life of less than a second. Kamen thought that the isotope pair nitrogen-14/carbon-14 would behave similarly to the helium/lithium pair. Nitrogen-14 was obviously stable. Carbon-14 would be unstable and, by analogy with helium-6/lithium-6, have a very short half-life. Kamen and his coworkers also thought that carbon-14 would, like other low-atomic-weight elements, emit an electron (beta radiation) when it decayed to nitrogen-14.

They got support for a short half-life from an analysis by Oppenheimer and Philip Morrison, who concluded that the decay from carbon-14 to nitrogen-14 would be quick, the half-life being on the order of minutes to hours. The reasoning that the half-life of carbon-14 was likely so short meant that whatever carbon-14 Kamen and Ruben could produce would not survive long enough to detect. Even though the understanding of the physics involved was rudimentary—that is, their calculations might be wrong—the presumed short half-life of carbon-14 meant that it was not a priority for time on the cyclotron. The search for carbon-14 was set aside.

Carbon-11, on the other hand, was relatively easy to produce and was one of the first created under the pharmaceutical and biomedical program for the cyclotron. Carbon-11 has a half-life of only 22 minutes. To make it, Kamen used diboron dioxide (B_2O_2), a powder, and bombarded it with deuterons, which you will recall are stable particles composed of a proton and a neutron.

Looking back, the initial use of carbon-11 seems almost counterintuitive, but it was in line with biomedical research at the time. The psychology department at Berkeley wanted to learn more about carbohydrate metabolism. Carbon assimilation through photosynthesis, about which nothing was known, was a side issue. The proposed experiment was to have plants assimilate carbon dioxide labeled with carbon-11 and then feed the now-radioactive plants to rats. Sam Ruben reluctantly went along with the psychology department program. He was annoyed at the whole prospect because he thought he had the original idea for using carbon-11 in rat metabolism. Given the less than half-hour half-life of carbon-11, the rat experiments never yielded any useful results. There was simply no way that the plants could produce enough food labeled with carbon-11 to feed to the rats, the rats then assimilate the carbon-11-labeled (what they thought would be) glucose, and the experimenters still have time for the assays. There were too many steps, and they all had to take place in the space of an hour or two—not to mention the problems of working with active animals as compared to unmoving plants.

Failing with the rats, Ruben had the idea to instead use carbon-11, at the first step in the experiment, to study the fixation of carbon dioxide in sugars through photosynthesis. Ruben, a chemist, originated and initiated the research into photosynthetic carbon fixation, even though it was a topic for biology. The failure of the rat experiments meant switching to what even then was recognized as the "big" problem, identifying the first product of carbon dioxide fixation in green plants. With a 22-minute half-life, the experiments were wild. Kamen would make a mad dash down the hill from the Radiation Laboratory, delivering a dose of carbon dioxide gas labeled with carbon-11 to Ruben's lab in the Rat House. Kamen would arrive at the door but go no further, because of his personal radioactivity.

I'll revisit the carbon-11 work in chapter 6, but unraveling the photo-synthetic carbon cycle was one spur that kept Kamen and Ruben work-ing together, eventually leading to the discovery of carbon-14. The other incentive came from Lawrence, whose funding sources were becoming skeptical of the potential for radioisotopes in medical research. Harold Urey, at a rival lab at Columbia University, was making the case that there were comparable stable isotopes, such as nitrogen-15, carbon-13, and oxygen-18, that could be used in medical research just as easily as their radioactive counterparts. Urey had earlier won a Nobel Prize (Chemistry, 1934) for the discovery of deuterium, a stable isotope of hydrogen. He also pioneered the separation of isotopes, a process useful to the Manhattan Project (development of the atomic bomb) during World War II. Now Urey was saying there was no need for radioisotopes. Stable isotopes could do everything and were available for the important elements of life: carbon, nitrogen, and oxygen, not to mention deuterium.

Lawrence sensed that if Urey's views prevailed, it would mean a loss of funding from foundations and chemical and pharmaceutical companies, and financial disaster for the Radiation Lab. He told Kamen and Ruben to determine, finally, whether there were long-lived radioactive isotopes in the top row of the periodic table, and for carbon, nitrogen, and oxygen.

Kamen and Ruben set themselves to the task, but they had competition of another kind. Requests for radioisotopes were becoming so great that the cyclotron operated around the clock, overriding the needs for basic research. Much of Kamen's work was relegated to the nighttime hours when cyclotron time for bombardments was possible. One by-product of this time scheduling was creating targets for the cyclotron that could be mounted internally. Internal targets could be used for research without interfering too much with radioisotope production using external tar-gets; the internal placement also meant Kamen's targets would be exposed to higher energies. According to Kamen, the change in target location became a turning point in the search.

■ ■ ■

As noted previously, Ruben and Kamen knew that carbon-14 existed, based on results from cloud chamber experiments years before. They could only guess, however, as to its chemical and physical characteristics except to say that it probably had a short half-life. There might be two ways of approaching it, using either a "bottom-up" or a "top-down" method. A bottom-up method would mean getting an extra neutron into carbon-13, the stable isotope of carbon. Carbon-13 exists at about 1 percent of the most common isotope, carbon-12. The idea was to bombard a sample of carbon-12, such as graphite, with high-energy deuterons and hope to get enough collisions with the minute quantities of carbon-13 to make carbon-14. The top-down approach would mean getting nitrogen-14 to give up a neutron, changing it to carbon-14. Kamen decided on the bottom-up approach because he assumed that carbon-13 atoms would present a larger cross-section, a larger target, to deuteron bombardment and therefore superior absorption by the graphite. Thus the plan was to bombard graphite, naturally containing small amounts of carbon-13, with deuterons. Although ultimately successful, it turned out to be the harder of the two methods.

Initially, they met with zero success. During the fall of 1939, Kamen, assisted by Emilio Segré and Robert Wilson, used alpha particles (helium nuclei) every way they knew how in the 60-inch cyclotron, now fitted with internal targets for greater energy exposure. They only produced prodigious amounts of carbon-11. Then they tried deuterons, the initial use of which also produced isotopes with short lifetimes. In January 1940, in parallel with the work on the 60-inch machine, Kamen set up a target internally in the 37-inch cyclotron. The target was designed to intercept any deuterons shot during the night. During the day, the target was removed, to leave unimpeded the regularly scheduled bombardments for the production of other radioisotopes.

Kamen made the target from graphite plastered onto a copper plate, knowing from his experience with cyclotrons that such a target could withstand the intensity of the deuterons from the cyclotron beam. He also upped the radiation. Instead of the previous 20 microamp hours of radiation, he increased the intensity to 5,700 microamp hours, equivalent to 10^{20} deuterons pounding the graphite. His routine was first to coat the copper plate with graphite and then, later in the morning, collect and count whatever had been "blasted off" overnight. Kamen's body had already absorbed enough radiation that he spiked background measurements near experiments, and the process of collecting the graphite fragments and inspecting the target each morning after bombardments meant even more hazardous exposure.

■ ■ ■

And so we return to that dark and stormy night, and the discovery of carbon-14. It was not an immediately obvious "eureka" moment. In the early hours of February 27, 1940, Kamen completed a bombardment, collected the graphite gravel, and put it into a sample bottle for Ruben's subsequent analysis. After his adventure with the police, Kamen caught up on sleep, not waking until the afternoon. Then he called Ruben, who had spent the day doing chemical tests on the sample. While Kamen slept, Ruben first burned (oxidized) the graphite gravel with cupric oxide to make carbon dioxide. Then he reacted the carbon dioxide gas that was produced with calcium hydroxide, precipitating calcium carbonate. Calcium carbonate is stable and could be easily kept, but to assay his sample in the counter, he needed the carbon as a gas. He reacted the calcium carbonate with sulfuric acid to make carbon dioxide again and passed the gas to his counter.

Kamen's gravel showed weak but reproducible radioactivity. By this time, Kamen had returned to the lab and watched, at a safe distance, as Ruben went through his purification cycles, acidifying the calcium carbonate to drive off carbon dioxide, then reprecipitating the gas to convert it back to calcium carbonate. Ruben went through this procedure several times, each time measuring the activity. The count, at a few hundred above background, was stable; the radioactivity was still there.

Interestingly, of all the elements in the periodic table, only carbon can be put through a cycle of acidification and precipitation under oxidizing conditions. That this is so bolstered their claim that the radioactivity they measured was an isotope of carbon. The counts remained stable, suggesting that carbon-14 was probably long-lived, at least longer than hours. Still, Kamen had concerns. What if they were measuring impurities in the graphite? For example, the cyclotron bombardments could have produced sulfur-35, an isotope with a long half-life, from sulfur impurities in the graphite. They performed a chemical test to precipitate any sulfur present to sulfate ion, and thereby separate it from volatilized carbon dioxide. Still, the counts remained.

Excited by these results, Kamen devised a new graphite target and produced much more isotope. He produced enough to show a radioactive decay of 100 disintegrations per second. He then calculated a rough half-life from three factors: how many deuterons were bombarding the target, the energy supplied by those deuterons to the cross section of the target, and the fact that carbon-13 comprised 1 percent of the carbon (as graphite). The half-life, the time it takes for half the radioactive isotope to decay, would be the natural logarithm of 2 (0.693) times the disintegration rate. So, he figured that with ~1.2×10^{20} deuterons he would make 6×10^6 carbon-14 nuclei. The calculation is $[(1.2 \times 10^{20} / 6 \times 10^6)/100] \times 0.693$, which worked out to be ~4,000 years! A long-lived isotope, indeed. We know now that carbon-14 has a half-life of ~5,700 years, but that is of small importance compared to the knowledge that science now had a long-lived radioactive isotope of carbon, carbon-14.

The first thing Kamen and Rubin did was to consult Gilbert Lewis, the best-known member of the Berkeley chemistry faculty, who merely expressed confidence in them: "If you guys say so, it must be true." Not entirely reassured, and balancing excitement with some doubt, they nevertheless rushed to Lawrence's home with the news. Lawrence was in bed nursing a cold. The next night he was to be awarded his Nobel Prize for developing the cyclotron, and he did not want to sneeze and cough his way through the ceremony. Perhaps because of the war, the Swedish consul in San Francisco was to present that year's prize at the Berkeley campus. In addition to his Nobel, Lawrence now got more good news. Ecstatic, forgetting his cold, he got out of his sickbed, danced around, and congratulated his young scientists, saying the discovery would be announced at the ceremony the next night. Lawrence's excitement did not dispel Kamen and Ruben's doubts, especially as the prospect of public disclosure seeped in.

The Nobel was awarded to Lawrence on February 29, just days after Kamen and Ruben's discovery. Kamen attended, but Ruben, losing his courage, stayed home. The chair of the physics department, R. T. Birge, opened the ceremony and spoke about the importance of radioactive isotopes to science and medical therapies. Then he said:

> I now . . . have the privilege of making a first announcement of very great importance. This news is less than twenty-four hours old and hence is real news. . . . Now, Dr. S. Ruben, instructor in Chemistry, and Dr. M. D. Kamen, research associate in the Radiation Laboratory, have found by means of the cyclotron, a radioactive form of carbon, probably of mass fourteen and average life of the order of magnitude of several years. On the basis of its potential usefulness, this is certainly much the most important radioactive substance that has yet been created.[4]

2

DISCOVERY'S WAKE

After Birge's announcement at the Nobel ceremony, Kamen worried that his fame would be short-lived if his and Ruben's discovery turned out not to be carbon-14. Then came the issue of authorship. Ruben only did the chemical identification of the carbon-14. Kamen did the bulk of the work, making the cyclotron run and devising the nuclear chemistry behind the discovery, no small tasks. But Ruben argued that he needed a first-author publication to get tenure in the highly competitive chemistry department at Berkeley, and he suspected (probably correctly) that he might otherwise fail. In the end, Ruben prevailed. And later, a longer contribution on which Kamen was to be first author also ended up as "Ruben and Kamen" because the secretary in the chemistry department, a force herself, switched the authorship just before sending it off. Kamen was outmaneuvered again. Lawrence was furious that his Radiation Laboratory was not given the primacy he thought due. For himself, Kamen felt that the authorship issue was his own problem, and continued his collaboration with Ruben until the war effort separated them. Sadly, Lawrence came to believe that Kamen's contribution to the discovery of carbon-14 was secondary.

Kamen and Ruben's discovery did not come without criticism. Was their sample really carbon? Could carbon-14 stay around for that long? Their isotope sample could be oxidized and made volatile from acid solutions. It could be precipitated and then revolatilized, repeatedly. As Kamen noted, carbon is the only element in the periodic table for which this is possible. Also in Ruben and Kamen's favor, later experiments showed that algae liked their product—the algae ate it up. Oppenheimer, however, believed that the decay of nitrogen-14 to carbon-14 would never result in a long-lived isotope, echoing some of Kamen's early concerns. Why was it not like the pair helium-6/lithium-6, nearby in the periodic table, in which one of the pair was stable and the other decayed in seconds? In the end, Kamen decided to believe the algae rather than prominent theoretical physicists. They stored their original sample in the Geiger-Müller counter and saw no real decline in activity after several months.

Carbon-14's long half-life, 4,000+ years, was a wholly unexpected discovery, and the reason it took as long as it did to discover. The researchers knew that an isotope of carbon with an atomic number of 14 could exist. They figured, however, that the half-life of carbon-14 would be on the order of seconds. If carbon-14 decayed over a longer period, say with a half-life of months, you would not need much to detect it; the decay rate for such an isotope would be rapid enough that only small amounts need be produced. But if the half-life were very long, the decay rate for each milligram of the isotope would be proportionately lower. The problem, initially, for Kamen and Ruben was that they never produced enough isotope to detect it decaying. And because of the exceptionally long half-life they found for carbon-14, even long after the discovery, there were those who still refused to believe that it was an isotope of carbon. As I'll show in chapter 6, the finding of carbon-14 in the first products of photosynthesis, a few years after World War II, proved at last that what Kamen and Ruben had discovered was indeed a long-lived radioactive isotope of carbon.

The beginning of World War II interrupted research using carbon-14. Even though the United States had not yet entered the war in Europe, the potential for war meant that all science and funds to support science were directed toward a nascent war effort. The Berkeley Radiation Laboratory was no exception. The government scrambled to assemble the scientific talent it felt it needed. Research that Kamen and Ruben hoped to complete, solving the riddle of the first products of photosynthesis, would have to wait. Kamen and Ruben were assigned different jobs but in their respective labs, Kamen staying in the Radiation Lab and Ruben staying in the Rat House. Both would be adversely affected by their wartime experience, Ruben fatally.

In the lead-up and during the War, Kamen continued producing iso-topes for various uses, but not carbon-14. On a war footing, everyone was under surveillance of some kind, leading to many misunderstandings. And for Kamen, the security measures were to have terrible consequences.

The government set up a few facilities around the country as private-public partnerships, and Kamen traveled to them for one reason or another, sometimes for reasons completely unknown to him. On one occasion, he was sent to Oak Ridge National Laboratory (ORNL) in Tennessee. While there, he inspected a laboratory where radioactive sodium was being produced. He was shown a container of some of the radioactive sodium, and when he peered inside, he saw a purple glow. He realized that the glow meant a much larger source of charged particles than what a cyclotron alone could produce. The glow could only mean one thing: there was a nuclear reactor somewhere at ORNL. Kamen, excited about this possibility, was not shy about letting people know (shyness was not one of his personality traits), including Lawrence, who was also visiting Oak Ridge. His loose talk about a nuclear reactor at Oak Ridge alerted the authorities, who believed that someone had given him the reactor information, and Kamen came under permanent and almost continual surveillance.

A year or so later, back in the Bay Area, he continued to play chamber music with Isaac Stern and others, and became acquainted with members of the local émigré population. At one event, he met a couple of Soviet consular personnel from the San Francisco consulate. They had heard about the therapeutic value of phosphorus-32, then being produced at Berkeley, where Lawrence was spearheading its use in the treatment of leukemia. They asked Kamen if he might relay a message to Lawrence requesting a sample for one of their consulate employees, who was suf-fering from the disease. Kamen was happy to comply, and when he spoke to him, Lawrence seemed happy to help.

By way of thanks, the two Soviets invited Kamen to dinner at Fisher-man's Wharf in San Francisco. Unbeknownst to him, he was still under surveillance, and being observed undercover at this dinner. Shortly there-after, Lawrence called Kamen into his office and summarily dismissed him from his position at the Berkeley Radiation Lab. He was out of a job, and never told why. Earlier, his wife, Esther, had announced her unhappiness with their marriage and moved out, filing for divorce. So, in early 1944, at 30 years old, Kamen was left with no means of support, financially or emotionally. And his loyalty to the United States was now suspect. Fortunately, his music sustained him, as did the friends he made through music. Through those friends, he managed to get a job working as an inspector at the Oakland shipyards. Through other friends, he was given permission to use a laboratory, remote from his old haunts, but still a working lab.

Eventually, again with help from friends, he found a position running the cyclotron lab at Washington University in St. Louis. The position was such that his research on photosynthesis, started with Sam Ruben, would not continue, even though by now carbon-14 was in ready supply. After the war, Lawrence selected Melvin Calvin to head a group to investigate photosynthesis. As described in chapter 6, Calvin's group, instead of Kamen and Ruben, discovered the first product of photosynthesis and the molecular acceptor of carbon dioxide, discoveries that led to Calvin's Nobel Prize.

In 1948, Kamen was dragged in front of the House Un-American Activities Committee to testify and, like J. Robert Oppenheimer, was dogged for years afterward by suspicion of being a communist spy. Oppenheimer, an outspoken opponent of nuclear proliferation, made political enemies during the time of the Red Scare witch hunts. Kamen was not political at all. Nevertheless, he was portrayed in the *Chicago Tribune* and the *Washington Times-Herald* (owned by the *Tribune*) as someone passing secrets to Soviet agents. Without explanation, the State Department held his passport application in limbo for years, denying him the ability to present his research at international conferences. One day, Beka, his second wife found him on the floor of their bathroom, bleeding from his wrists and throat—an attempted suicide. In 1955, he sued the *Tribune* for libel and the State Department for violating his right as a citizen to obtain a passport.

Ironically, during the trial for libel, he was able to see the file the FBI had on him, and was surprised to learn that no record existed of the dinner with the Soviet consular personnel— the event that had cost him his job at the Radiation Lab. One invitation to speak at a conference in

Argentina illustrated the silliness of the denial of his passport application during those years. Kamen had a cousin (formerly unknown to him) who was high in the administration of Argentina's president, Juan Perón, hardly a communist sympathizer. He related the invitation directly to the State Department, which must have been astonished to find that an alleged communist sympathizer would be a guest of the Perón government. After the trial, the State Department reluctantly issued him a passport, and the court awarded him damages for the libel. After a long legal struggle, he claimed victory over both the *Tribune* and the State Department.

Finally exonerated after so many years, Kamen spent his last decades working on problems of cellular metabolism, particularly involving chromatophores and cytochromes. On the side, he helped establish the biochemistry department at Brandeis University, and later assisted Roger Revelle in creating a new University of California campus in La Jolla, which became UC San Diego. In 1996, he was presented with the Enrico Fermi Award for lifetime achievement in energy research. The award came from the Department of Energy, who now operated very same the lab that fired him in 1943. He died in 2002 at the age of 89 in Montecito, California. Exposure to radioactivity during his early years as a nuclear chemist seems not to have affected his longevity.

One story in Kamen's career echoes today, at least among limnologists and oceanographers like myself. At a conference at the University of Wisconsin–Madison, he learned about cyanobacteria, formerly known as blue-green algae. On occasion, these bacteria bloomed to smelly, turbid proportions in Lake Mendota, adjacent the UW campus, because of excess phosphates from wastewater. A plant physiologist there, Folke Skoog, studying these blooms, recommended dumping iron salts into the lake to precipitate the phosphate, much as water treatment plants in cities do today to remove any mobilized lead from old pipes. The result was opposite to the intended effect: the cyanobacteria bloomed as never before. This might have been, unwittingly, the first iron-enrichment experiment in aquatic systems, preceding the famous ones in oceanography 40 years later.[1]

World War II also interrupted the life of Sam Ruben. During his days at Berkeley, Ruben was a chemist held in very high esteem, an experimentalist without peer, offering great promise. Kamen said that it was his expert planning that was responsible for much of the Radiation Lab's early successes. Working with carbon-11, with its half-life of only 22 minutes, certainly demanded expert planning.

Ruben died, tragically, in an accident involving chemical weapons. War mobilization involved everybody, and Ruben's job was to understand the physiological effects of phosgene gas, a chemical weapon, to devise an antidote to exposure. This meant exposing rats to phosgene gas, then sacrificing them to see how it behaved in their bodies. In addition to this project, he had been working on a related one: how these kinds of gases might spread in the environment. He did the work at beaches in Marin County, and at Mount Shasta in northern California.

The project had him working around the clock for several days. Driving back to Berkeley, he fell asleep at the wheel, drove off the road, and broke his right wrist. He returned to work after a weekend's rest at home, his arm in a sling. Back at the lab, Ruben found that the phosgene supply he had used for the rat experiments was finished. He needed to transfer from the original phosgene supply, stored in soft glass ampoules, to a safer steel "bomb." Instead of immersing the ampoule in ice salt, which would have meant waiting, he tried to open the ampoule in liquid nitrogen, so that it could enter a vacuum line. Even though his arm was in a sling, he refused help from assistants, concerned for their safety. Unfortunately, the glass of the phosgene ampoule was flawed and cracked during the tricky operation, spraying a fine mist of phosgene gas into the lab air. Ruben was in the direct line of exposure. He calmly sat down, and called the hospital. The nature of the lethal gas, he knew, meant that he had to remain immobilized. He was transported to the hospital, but too late. A massive pulmonary edema developed, and he died that night. He left a wife and three children.[2] It was said that in addition to the war effort, Ruben felt pressure to get tenure, and also put pressure on himself to succeed. And maybe the strain of work was one reason he never turned in the forms for federal insurance compensation, compounding the tragedy. Science lost a gifted experimentalist, and one of the two discoverers of carbon-14.

Many of the others that Kamen and Ruben worked with at the Berkeley Radiation Laboratory went on to their own success, becoming a sort of "who's who" in twentieth-century nuclear physics and chemistry. Robert

Wilson, Ed McMillan, and Emilio Segré all joined Oppenheimer at the Manhattan Project's Los Alamos laboratory during the war. McMillan and Segré both received Nobel prizes, in chemistry and physics, respectively (see appendix 1). McMillan returned to Berkeley and shared in the invention of the synchrotron, a cyclotron with increased energy output in the particles produced. Robert Wilson later founded and directed Fermilab in Batavia, Illinois, and after positions at various major university physics departments, retired to Ithaca, New York.

There is one further story regarding the discovery of carbon-14 that needs to be recounted because of its later repercussions. A. T. Birge, the chair of the physics department, in announcing the news about carbon-14 at the Nobel ceremony, said that the cyclotron "created" the isotope. Kamen's discovery, helped by Ruben's chemistry, was just that, a discovery. Unbeknownst to them, nature had been producing carbon-14 all along—from nitrogen.

Instead of using nitrogen, Kamen and Ruben created carbon-14 by bombarding graphite with a high enough energy intensity and for a long enough time to transform the 1 percent of carbon-13 naturally occurring in the graphite to carbon-14. Here, as I said, they were going "bottom-up" (carbon-13 to carbon-14) rather than "top-down" (nitrogen-14 to carbon-14). Even though the method succeeded, by brute force, Kamen soon realized that this route to carbon-14 would not be practical. Bombarding graphite with deuterons would never create enough isotope for biomedical purposes, for example. Fortunately, Kamen was a smart enough scientist to look at the problem from different perspectives.

Neutrons coming out of the cyclotron were never accurately parallel; they were scattered about more or less randomly and, for all practical purposes, lost. Kamen thought he might be able to take advantage here, and employ these otherwise wasted neutrons. Along with their intensive search with deuterons and graphite, in January 1940, a month before the actual discovery, Kamen designed a side experiment. He filled two five-gallon carboys with a solution of ammonium nitrate, NH_4NO_3, and then, with Ruben, acidified them[3] and purged them of carbon dioxide gas. This last step is interesting. The idea was to remove the carbon dioxide that might dilute any carbon-14 produced, not that they expected to find any. That attention to detail would prove prudent.

He placed the carboys around the periphery of the 60-inch cyclotron beam to capture the scattered neutrons. After a while, however, Kamen forgot about the carboys, focused as he was on his graphite targets used with the old 37-inch machine. He only remembered the carboys when the cyclotron technicians complained. They were the ones who had to haul them out of the way to make adjustments to that part of the system and then muscle them back into place. Then the carboys sprang leaks, and that was the last straw—they had to go. Kamen took them to the Rat House lab, where Ruben went through his usual procedure to capture any carbon. Surprisingly, a large amount of carbon could be precipitated out of the acid solution, which they thought meant that carbon dioxide from the local air was diffusing back into the carboys. But when they assayed the precipitate, the counter pegged. They had, almost unwittingly, produced more carbon-14 than had ever been created, several microcuries. This put an end to any further interest in using carbon-13 and graphite, the bottom-up approach. Using neutrons to bombard nitrogen-14 is how carbon-14 has been produced ever since.

Kamen now remembered an earlier missed opportunity. A few years earlier, in 1936, when physicists and chemists were only dimly aware that a radioactive isotope of carbon might exist, they recognized the pathway of nitrogen-14 giving up a neutron to become carbon-14. When Kamen first arrived at the Berkeley Radiation Lab, Ed McMillan filled a small bottle with purified ammonium nitrate and put the bottle in the path of neutrons from the cyclotron. The bottle was left there for a few months while everyone attended to other experiments. One day, the bottle was found broken, its contents spilled. The ammonium nitrate was never analyzed, but many years later, Kamen realized that the bombardment had probably created significant carbon-14.

As if the distraction of the war were not enough, Lawrence doomed a follow-up to Kamen and Ruben's serendipity with the carboys. Lawrence believed that the ammonium nitrate would explode, and no amount of explanation, entreaties, or counterarguments from Kamen could budge Lawrence from his stance. In Lawrence's mind, an explosion, however remote the possibility, would mean total catastrophe. Nevertheless, Kamen and Ruben realized that making carbon-14 from nitrogen was much more efficient than transforming the tiny amount of carbon-13 in graphite. The atmosphere, 80 percent N_2, abounds with nitrogen. Is it possible that some is converted to carbon-14 by some energetic process? Stay tuned—that is the topic of chapter 4.

■ ■ ■

The story of the discovery of carbon-14 has many threads, and I have spent time and the reader's indulgence to tell it. But why is a long-lived radioactive isotope of carbon so important? Carbon defines life on Earth; all life is based on carbon. It is probably just as important to ask, why is life based on carbon?

First, a look at the carbon atom itself reveals unique properties. The carbon atom can combine in four atomic locations, ideal for combining with other atoms or with itself, thus forming other molecules. The carbon nucleus has six protons and six neutrons, giving it an atomic number of 6 (its place in the periodic table of elements) and an atomic weight of 12 (figure 2.1) Just outside the nucleus, two electrons orbit the nucleus, and outside of that orbit, there are another four. Chemists call these orbital

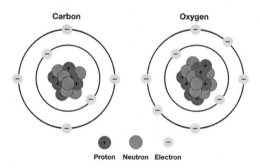

FIGURE 2.1

The carbon atom (*left*), showing the nucleus (protons and neutrons) and the two electron "shells": the filled inner shell (two electrons) and the partially filled outer shell (four electrons). According to atomic theory, the outer shell can have up to eight electrons. The oxygen atom (*right*), showing its nucleus (protons and neutrons) and two electron shells, the inner one with two electrons, and thereby filled, and the outer shell, partially filled, with six electrons.

arrangements "shells" that, according to atomic theory, have a prescribed number: two are allowed to fill the inner shell and eight allowed in the next outer shell. An atom with a nucleus of two protons and two neutrons is helium, atomic number 2, an inert, chemically unreactive gas with its inner shell of electrons filled. Filling the next outer shell means an atomic number of 10, which is neon, also an inert, unreactive gas. For carbon, two electrons fill the inner shell and four are in the outer shell.

Atomic theory says that the carbon atom would need another four electrons to fill that outer shell, giving it the right number, theoretically, and making it chemically stable. Instead, carbon tends to either give up the four electrons or acquire four others by sharing with other atoms. Chemists call this a "valence" of 4. Sharing electrons describes the nature of chemical bonds between atoms, thereby forming molecules, and is used by carbon and other atoms to achieve stability.

The oxygen atom, at an atomic number of 8, two positions along from carbon in the periodic table, has two missing electrons in its outer shell, for a valence of 2. Two atoms of oxygen, each offering two of the electrons that carbon needs, make carbon dioxide (CO_2), with a double bond of carbon binding each oxygen atom. The carbon's electron shell is now filled, and the carbon is said to be "fully oxidized" by donating its valence electrons, forming a double bond (figure 2.2).

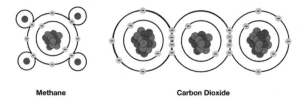

Methane   Carbon Dioxide

FIGURE 2.2

The methane (*left*) and carbon dioxide (*right*) molecules, showing how the electrons are shared between the two oxygen atoms, or the four hydrogen atoms, and the one carbon atom. Carbon dioxide is said to be "fully oxidized," by virtue of sharing electrons between the carbon and oxygen atoms. Methane is said to be "fully reduced," by sharing electrons with four hydrogen atoms. In both cases, carbon's outer shell is filled.

At the other extreme is hydrogen, which can offer one electron to share. Four hydrogen atoms can bond with one carbon atom to make methane (CH_4), filling the outer shell of the carbon atom. Hydrogen's giving the electrons to the carbon atom in this way means that the carbon atom is "fully reduced." So carbon can combine with itself, with hydrogen (by far the most abundant element in the universe), and with oxygen (the most reactive) to create millions upon millions of what are called "organic" molecules—that is, molecules containing carbon. Thus, carbon can exist as fully oxidized (CO_2), as fully reduced (CH_4), or in any state in between.

These oxidation and reduction possibilities account for the ability of carbon molecules to exist on Earth as a solid, a liquid, or a gas. We are all familiar with the major atmospheric greenhouse contributor, carbon dioxide, an odorless gas. It is what we exhale from our lungs as a product of respiration in our bodies. It is expelled in respiration from almost all living cells, is produced from burning (oxidizing) organic matter, and is what plants and algae use in photosynthesis. Carbon can also exist as a liquid, as any trip to a gasoline station to fill up your car's tank will show. Oils, whether extra virgin olive or Kuwaiti crude, are long chains of carbon and hydrogen, methane molecules strung together—hydrocarbons— and once living. In solid form, carbon exists as charcoal, trees, rocks such as limestone, oyster shells, fibers, and insect exoskeletons.

Carbon's ability to exist as a solid, a liquid, or a gas at everyday temperatures means that living cells can store, expel, transform, or deliver carbon molecules within and across membranes. Carbon passes through cellular membranes either as a liquid or as a gas dissolved in cellular cytoplasm. In this way, it can communicate its chemistry among cells, or react in cellular chemistry. Water is the other major feature of life, and the solid, liquid, and gaseous forms of carbon mean that it can be soluble (e.g., carbon dioxide, sugars) or insoluble (oils, wood) in water. Variations in solubility allow for biological storage and structure (again, oils, wood) or exchange (sugars) when the solids are chemically changed.

In their simplest form, organic molecules are alcohols, aldehydes, sugars, fatty acids, amino acids, and nucleic acids. These simple molecules build into more complex molecules, the stuff of life. Sugars become carbohydrates and the more complex polysaccharides. Fatty acids become lipids—that is, fats. Nucleic acid molecules form chains to become deoxyribonucleic acid (DNA) or ribonucleic acid (RNA), the molecules encoding genetic information, the makeup of genes. Silicon also has a valence of 4 and, like carbon, can bind with itself to create long chains and even sheets. But silicon cannot exist as a liquid or gas, only as a solid.

Finally, life defined by carbon involves complexity—the formation of big, or macromolecules. If the millions upon millions of organic molecules and chemicals are not enough, carbon chemistry allows folding of the molecules to create three-dimensional structures. Amino acids link together in a chain to create peptides, and peptides link together, and fold, to create three-dimensional proteins. The three-dimensional molecules allow for unlimited intricacy, and they can house, within their folds, metal ions that allow them to transfer electrons and energy. Hemoglobin, the protein in mammalian blood, is one such macromolecule, its red color coming from an iron atom held at the molecule's center.

That is a summary of why carbon is the element of life, the biosphere. What is the importance of having a long-lived isotope of carbon? Once carbon-14 was created in the cyclotron, it was going to be there for a while, available for biochemical research. The chemistry of photosynthesis will be the subject of chapter 6, to show how carbon-14 has been used to reveal the pathways of carbon in photosynthesis and in other metabolic cycles. Research into biochemical cycles was otherwise completely stalled before the discovery of carbon-14.

Thus, we have the third ingredient to the importance of carbon-14 in our lives. First, a radioactive isotope of carbon, the major element of all life, was found to exist. Second was the unexpected finding that carbon-14 has a long half-life. Third, as I discuss in chapter 4, carbon-14 is being continuously created in the atmosphere from cosmic rays. Cosmic rays interacting with atmospheric molecules create neutrons that, like the particle beams of the cyclotron, transform nitrogen-14 to carbon-14. Carbon-14 quickly oxidizes to carbon dioxide. With solar energy—sunlight—carbon dioxide combines with water in photosynthesis, creating organic matter. We, and other animals, the heterotrophs, eat the plants. As a result, we are all radioactive. We have an intimate relationship with carbon-14. One could call even call it a "love affair." Carbon-14 is the translator for all that carbon can be: the tag, the message, for what organic molecules are, their history, and how they act.

Martin Kamen could only know he was radioactive because of specialized instruments. He showed no outward signs, and went about his life normally. His condition was invisible to all, including himself. Next, we'll look at that "invisible phenomenon," radioactivity, and how it is detected and assayed.

3

THE "INVISIBLE PHENOMENON"

Henri Becquerel wrapped his rocks in a black cloth and left them sitting on a photographic plate protected by black paper. He put his experimental setup in a drawer for the night, an additional safeguard to keep out any stray light. He hoped that perhaps tomorrow, after three days of cloudy weather, it would be sunny, and he could complete his experiment on the phosphorescence of his rocks. The year was 1896.

By now, most people are familiar with how film photography works, even though it has been quickly and almost totally taken over by digital imaging.[1] Energy, in this case light, chemically reduces light-sensitive salts, such as silver bromide, coating a paper or glass plate. The degree of chemical reduction and darkening depends on how much energy hits the coating.

Photography has been around since the early to mid-1800s, when it was practiced most commonly in France. One early practitioner, Abel Niépce de Saint-Victor, noticed that uranium salts could darken photographic coatings with an "invisible" light or radiation. He thought that it might be a delayed reaction in the dark from previous exposure to sunlight. Edmond Becquerel reported this finding in a monograph, and Becquerel's son, Henri, also a physicist, took up the work on the phosphorescence of minerals.

Henri Becquerel was also influenced by the excitement surrounding the discovery of X-rays the previous year, by William Roentgen in Germany.[2] People marveled at the images of their bones on phosphorescent screens. Becquerel thought that he might find X-rays from the phosphorescence darkening his photographic plate. His experiments began where Niépce de Saint-Victor's left off. He used slabs of a uranium mineral, uranyl sulfate, sandwiching the mineral with a photographic plate protected from the light. The idea was that when placed in the sun, the phosphorescent mineral would translate the sun's energy and produce an image on the plate, the hoped-for X-rays penetrating through the black paper covering. The phosphorescence Becquerel was looking for resembles the mineral used today in wristwatch faces, getting charged in sunlight during the day so that the watch numerals can glow for a time in the dark.

Becquerel took his rocks sitting on the photographic plate from the drawer the next morning, but, regrettably, it was to be another cloudy day. Instead of waiting yet one more day, he removed the black cloth and developed the photographic plate, expecting to find only weak images. Instead, the images had the same intensity as he had found in earlier experiments with the mineral exposed to full sunlight. For this mineral, the image on the photographic plate could be created without exposure to sunlight. Becquerel was at a loss to explain this "new class of phenomena." Subsequent experiments with uranium salts that do not phosphoresce showed the same ability to darken photographic emulsions. No light was needed to produce the effect; the energy came from the mineral itself.

Becquerel's published work on his experiments did not receive much attention, except from a Polish student living in Paris, Marie Sklodowska, who was married to Pierre Curie. For her doctoral dissertation, Curie (née Sklodowska) searched for new sources of Becquerel's "invisible energy" and, in so doing, came up with the name "radioactivity." Becquerel, along with Pierre and Marie Curie, shared one of the first Nobel Prizes in Physics, in 1903, for discovering "spontaneous radioactivity." Marie Curie was the first woman to win a Nobel Prize. In the course of her searches for radioactive substances, she discovered and named the elements polonium (honoring her native Poland) and radium, which contributed to her Nobel Prize in Chemistry a decade after her first Nobel in Physics. Early in the twentieth century, the dangers of radioactivity were unknown, and it is possible that Marie Curie's death (in 1934) was the result of long-term exposure to the pieces of radium she carried around in her pockets and also kept in her desk. Her illness may also have come from her work during World War I caring for the injured. During the war, she set up mobile radiological units for X-rays and used the gas emanating from radium (later identified as radon) for sterilizing wounds.

Today, Becquerel's name is used for the standard unit of radiation, the becquerel, representing one radioactive disintegration per second. For most of us, however, the curie is a more convenient if less universal measure. The curie, honoring its namesake, is based on the radioactivity of radium, 3.7×10^{10} disintegrations per second. For carbon-14, a microcurie is 37,000 disintegrations per second, and is the standard unit for commercial preparations.

We know now that radioactivity is the result of unstable atomic nuclei—unstable either through size or because of being bombarded by other kinds of radiative energy. After a certain atomic number in the periodic table (see appendix 2), the atoms of many elements become unmanageable. Viewed from the nucleus, it's like trying to corral an ever-widening pasture of horses. The ones closest in are easier to control, but the task becomes more difficult with increasing distance from the barn. So, these elements with high atomic numbers can easily fall apart—they disintegrate and change into something else.

In assessing radioactivity, any early investigations had to be accomplished by using photographic plates. To derive anything quantitative from this method quickly becomes tedious, but Becquerel's original method has had staying power. Investigators in the 1940s used photographic plates to find the early products of photosynthesis and to work out how carbon dioxide becomes incorporated into, and grows, organic matter. (I discuss the chemistry of photosynthesis in chapter 6.) I remember as a graduate student wearing "film badges" attached to the belt on my jeans as a way of assessing radiation exposure. In those days, no matter the isotope or emitter, a film badge had to be worn by everyone working in a radiation lab, nuclear power plant, or similar facility. Combined with various kinds of filters, the film badge can discriminate among the different types of radiation and record a relative dose. For those of us working with carbon-14, it was, frankly, useless. Carbon-14 would never expose a film badge, but such were the safety requirements for anyone working with any kind of nuclear radiation.

Soon after the Nobel was awarded to Henri Becquerel and Marie and Pierre Curie, Hans Geiger coinvented a means of detecting radiation more quantitatively, and used it in Ernest Rutherford's Cavendish lab in Cambridge. (Geiger is also credited with the discovery of the atomic nucleus.) One thing that puzzled early researchers in radioactivity was that, unlike phosphorescence, the energy emitted from a substance never changed no matter how it was treated or abused. Heating, cooling, sealing samples away—nothing seemed to alter the energy release. Later, however, changes were noted in some of them, the radioactive energy declining with time. Rutherford, among other discoveries, came up with the idea of a radioactive element's "half-life," used to differentiate the various radioactive isotopes, and a concept used in radiometric dating (the subject of chapter 4).

Radioactive decay, producing the energetic particles, is a random process. Each decay comes about by chance. But with enough radioactive atoms, the rate of decay can be quantified, and becomes a useful property. As defined by Rutherford, the half-life is the time it takes for an isotope's radioactivity to decrease by one half. Half-lives can range from minutes to millions or even billions of years. Some elements are so unstable that, while they might have been created in the elemental factories of stars, they no longer exist in nature. The element technetium is one example, as mentioned in chapter 1. Its discovery in a piece of cyclotron shielding opened the door to using Berkeley Radiation Lab's cyclotron for creating those otherwise absent elements. Also remember that for carbon-14 there was the catch-22 regarding its half-life. Martin Kamen and his colleagues assumed a very short half-life for carbon-14, so they could explain away their lack of success as not creating enough isotope for it to be around long enough before it decayed. In fact, no decays were detected because the half-life was too *long*.

For our purposes, particles (or photons) of radioactivity come in three basic types. Perhaps for simplicity, Rutherford assigned to them the first three letters of the Greek alphabet: alpha, beta, and gamma. There are about as many permutations of this basic classification as there are elements in the periodic table, which we need not get into. In the simplest scheme, alpha particles are the nuclei of helium, element number 2 in the periodic table. Helium (He) has two protons and two neutrons bound together in its nucleus, and that is the alpha particle. Beta particles are electrons, and gamma rays are photons with high energy.[3]

Carbon-14 is what is known as a "weak beta emitter." Its radiation does not travel very far, and beta particles will be stopped by, for example, aluminum foil. As an illustration of how weak its radiation is, carbon-14 is sometimes used in paper mills to gauge the thickness of paper after it has been produced. Different amounts of radiation will be absorbed by the paper depending on its thickness. So even paper can protect pretty well against the radiation produced by carbon-14. Of course, you wouldn't want to drink a solution of an organic solvent labeled significantly with carbon-14, because there is the potential of its becoming part of an organ or muscle, in which case the carbon-14 would be putting out that beta radiation until you're dead, and for a long time thereafter. Drinking many types of an unlabeled organic solvents (acetone, methanol) would cause far greater and more immediate problems than ingesting carbon-14.

■ ■ ■

Some of us remember the movies and TV dramas from the 1950s, the so-called dawn of the nuclear age, when nuclear disaster films had a heyday—not that disaster films ever really left popular culture. Film directors needed a device to make visible and audible the invisible and inaudible danger of radioactive people or things, or nuclear bomb fallout. That film device, and the real-world one, was the Geiger counter, clutched by the hero, giving off its ominous ticks, and becoming a continuous and loud static as danger approached.

On location for one ill-fated movie, John Wayne, decked out like Geng-his Khan, pulled a device out of his bag that would have been unrecognizable to a thirteenth-century Mongol warlord, and turned it on. At first he thought the crackle meant a defect, but then he realized his Geiger counter was recording a high level of radioactivity. What prompted Wayne to bring a Geiger counter to the movie location in southern Utah is anybody's guess. The movie was *The Conqueror* (figure 3.1), one of many epics from Hollywood in the late 1950s and early 1960s; produced by Howard Hughes, it debuted in movie theaters in February 1956, but location filming took place in 1953.. The Atomic Energy Commission, in the midst of atomic weapon tests, had assured Hughes that southern Utah, the location for *The Conqueror* was safe because it was more than 100 miles from the

FIGURE 3.1

Movie poster advertising *The Conqueror*. (Wikimedia Commons)

test site at Yucca Flats, Nevada. It was indeed 100 miles away, but it was also downwind of the test site. The actors, crew, and a multitude of extras endured daytime temperatures reaching 120°F (49°C). Hot winds and "Mongols" on horseback continually kicked up dry dust. It was this dust, swirling around the movie battle scenes, that set off Wayne's Geiger counter in teaming ticks. Hughes, believing the Atomic Energy Commission when they told him the area was safe, and perhaps trying for some verisimilitude, brought 60 tons of the dust back to Hollywood to finish the film on a studio lot, possibly continuing to contaminate the actors and crew after the southern Utah filming.

Ninety-one of the 200 or so actors and crew were stricken with cancer, including Wayne, his costar Susan Hayward, and the director, Dick Powell. Forty-seven died, including Wayne, who succumbed to stomach cancer 25 years later. Strangely, even though he used a Geiger counter, Wayne was reportedly unconcerned about the dangers of radiation. That seems to have been a common view at the time.[4] Attributing cancer deaths to radiation exposure is difficult, especially when other factors (Wayne and Hayward were both smokers) enter into the picture. Still, it has been asked, "Did the U.S. government kill John Wayne?"

Hans Geiger invented his namesake counter in 1908. The original device could detect only alpha particles, although other experimental methods at the time could detect beta radiation as well. In 1928, Geiger's student Walther Müller added a sealed tube, and the Geiger-Müller detector was born. The new device could measure all three radiation types—alpha, beta, and gamma emissions—depending on the individual design. Müller's modification to the original was simple enough that the counter could be mass-produced. Its power consumption was low, and the electronics simple. The Geiger-Müller detector initiated a long era of portable radiation counters, with few modifications since its invention.

Today's Geiger (or Geiger-Müller) counter consists of a tube containing a noble (inert) gas, such as neon or helium, and a "window" on the end (figure 3.2). In the most common configuration, the window is made from the mineral mica, 1–2 millimeters thick. For this version, the mica window is thin enough to allow beta particles, but not alpha, to enter the tube. A wire is situated along the central axis of the tube and extends the length of the tube. A high voltage is applied, typically 400–600 V. When a beta particle, or any other high-energy particle, passes through the window and enters the tube, it collides and interacts with molecules of the gas. This interaction ionizes—that is, charges—the gas molecules, making them conductive. The resulting electric signal is picked up by the wire and recorded and displayed, usually as a "count" representing a radioactive decay event. A small speaker transforms the electrical signal into that

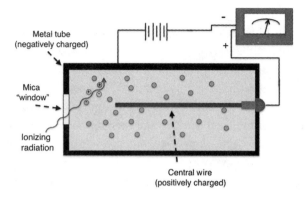

Metal tube (negatively charged)

Mica "window"

Ionizing radiation

Central wire (positively charged)

FIGURE 3.2

Schematic of a Geiger-Müller tube, showing the basic components. Alpha or beta particles enter the tube through the mica window. (Gamma radiation enters through the metal walls of the tube.) The tube contains an inert gas, such as neon or argon, which becomes ionized from the interaction with the radioactive particles (or electrons excited by gamma rays). Once the gas is ionized, the circuit between the cathode (central wire) and the anode (metal tube) is complete, and the electrical signal, a voltage pulse, is proportional to the number of particles ionizing the gas. The pulses are measured and processed, and recorded as "counts."

familiar "tick" or "chirp." Gamma rays can also be detected in modern Geiger counters, but in a different way. At their high energy, gamma rays pass through the wall of the counter, in the front, back, or along the sides, and not through the window. The wall is constructed from thin chrome steel. The gamma rays produce electrons in the steel wall, which enter the tube to ionize the gas molecules. Depending on how they are used, Geiger counters can be optimized for particular isotopes; some are used exclusively for carbon-14.

Geiger-Müller counters can be made relatively cheaply, and found wide use in the earlier decades of the twentieth century. They were great detectors of overall radioactivity, but in laboratory experiments, they proved cumbersome and limited the number of samples that could be analyzed. When I started at the Lamont-Doherty Geological Observatory in the late 1970s, one whole lab in the marine biology wing of the building was devoted to a Geiger-Müller system. No one seemed to know where it came from or what to do with this massive piece of outdated analytical equipment. Most of the mass, if not the size, came from lead bricks stacked around the detector and used to shield it, and the samples to be analyzed, from environmental (and therefore spurious) radiation (see chapter 4). When radioactivity is measured in decays numbering in the few hundreds or thousands, naturally occurring radiation can make up a significant part of the total. I called the manufacturer, amazingly still in existence, to find out if the system was worth anything. The representative replied, "Well, you've got lots of virgin-lead bricks that have value." Over the years we used those bricks as supports for laboratory setups and, more typically, as doorstops.

■ ■ ■

"If you want to stay safe, and maybe live long, don't live in Denver, and don't stand in line at Radio City Music Hall." That was the first thing Phil Lorio said to our radiation safety class at Lamont-Doherty. Lorio was Columbia's very first radiation safety officer, which tells you something about how old he might have been at the time, how recent radiation safety work was then, or how old I am. He was part of the Manhattan Project at Columbia during World War II and stayed at Columbia, working out of the School of Engineering.

Phil came out every once in a while from Morningside Heights, Columbia University's neighborhood in Manhattan, to Palisades, New York, the home of Lamont-Doherty, to see how we were doing, check our procedures, and give his class. Attending his radiation safety lecture meant that one could become a "radiation worker" or "radiation supervisor." Airline pilots and flight attendants (see the next chapter) are also classified as radiation workers in the United States and many other countries. I learned that those of us who were using carbon-14 were in the minority, and given the radioisotope's characteristics, we could be easily disregarded. Procedures for carbon-14 could be included in normal laboratory safety precautions for handling chemicals and solvents. Most in attendance at these sessions were using X-ray machines for analyzing and visualizing deep-sea sediment cores, with exposure for them being much more dangerous than exposure for us. But anyone working with any kind of radioactivity or ionizing radiation sources was required to attend Phil's safety lectures. So why Colorado? Why Radio City Music Hall?

Before I answer those questions, let's look at radiation in general, and ionizing (the dangerous kind) in particular. Radiation, as represented by the electromagnetic spectrum, goes from gamma rays and X-rays at the short-wavelength, high-frequency, end to the infrared, radio waves, and microwaves at the long, low-frequency end (figure 3.3). In the middle, at wavelengths of 400–700 nanometers (10^{-9} meters), is where our eyes sense radiation, the visible part of the spectrum, the world shown to us through our retinas. Radiation can also be divided into those wavelengths that are ionizing and those that are nonionizing. The visible (400–700 nanometers) and longer wavelengths, generally longer than the ultraviolet, are nonionizing.

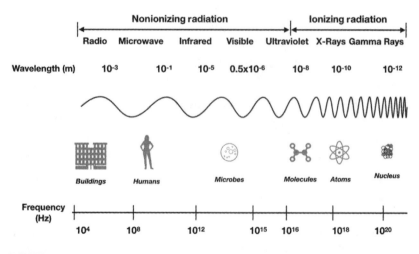

FIGURE 3.3

The electromagnetic spectrum. The wavelengths are shown to correspond to the approximate sizes of humans, insects, bacteria, and so forth. The dangerous radiation (ultraviolet, X-ray, and gamma ray) is at the shorter-wavelength, higher-frequency end of the spectrum. The visible portion of the spectrum, that which our eyes can sense, is between 0.4 and 0.7 micrometers, or 400–700 nanometers. At the long-wavelength end of the spectrum, we sense infrared and microwave radiation as heat, and radio waves are used to transmit signals over long distances.

"Ionizing" means that the energy in the radiation is strong enough to free electrons from molecules, making them ions, or charged particles. Radioactive decay, as I mentioned briefly in chapter 1, is a source of ionizing radiation. Also in chapter 1, I discussed artificial means of producing radioactivity, such as with the cyclotron. Radioactive decay produces alpha and beta particles and gamma rays, all of which are ionizing, able to free electrons from molecules. As Henri Becquerel stated, radioactivity is "invisible."

Ionizing radiation strikes the Earth all the time, and radioactive decay occurs every second. Both of these contribute to what is known as the "background" radioactivity on Earth. We cannot avoid it, unless, like some of the samples we wish to assay for radioactivity, we live inside a chamber made of lead bricks. Carbon-14 decays are beta particles, electrons, and are not particularly dangerous, even at the high levels used in biochemistry and ocean-productivity experiments, discussed later. Often, carbon-14 is added as labeled carbon dioxide, which for us heterotrophs is a waste product of metabolism anyway.

Lorio's point was that radiation is everywhere, certainly in the upper atmosphere and outer space, and also in some unexpected places. Colorado has mountains; Denver is the "Mile High City." The closer you are to the top of the atmosphere, the more exposed you are to cosmic radiation (cosmic rays), which, like radioactivity, is ionizing radiation. The FAA recommends limits on the number of hours that airline pilots and flight attendants spend at or above 30,000 feet for the same reason, thereby curbing their exposure to cosmic radiation.

At Radio City Music Hall, radiation comes from the rock used to build it. That rock contains small amounts of radium, a radioactive element discovered and named by Marie Curie. Radium decays to radon, an inert gas, emitting a gamma ray in the process. Standing in line next to the building means slightly greater exposure to radiation from the decay of radium to radon. The name "Radio City" takes on a new significance when you realize what is in the rock. (Actually, higher radiation levels can be recorded at a nearby tourist attraction and transportation hub, Grand Central Terminal.) Phil Lorio's were tongue-in-cheek warnings. People live comfortably in the Rocky Mountains or stand in line, maybe uncomfortably, for a Rockettes show. The point is, radiation is everywhere, a bit higher in some places than in others.

Is living in Colorado or flying jetliners ultimately harmful? Pretty much everything in excess harms health; even consuming too much water can produce deadly ionic imbalances in our bodies. Our environment presents a wide variety of assaults and insults, and life has a wide variety of responses, adaptations, and correctives. Ionizing radiation can be dangerous all the same. It can destroy living tissue, cause cancer, or lead to radiation sickness. The other side of the coin is using radioactive isotopes for the treatment of cancer. The production of isotopes for cancer treatment funded the Berkeley Radiation Laboratory and Lawrence's cyclotron in the early days, before the existence of federally funded scientific and biomedical research. Iodine-131 can be used to treat thyroid cancer, and strontium-89 is used to treat bone cancer. Instead of affecting healthy tissue, the idea is to destroy the cancerous tumor. As described in chapter 1, Martin Kamen unwittingly invited trouble and scrutiny when he served as a go-between for Soviet consular officials seeking radioisotopic treatments from Lawrence and the Radiation Laboratory to help one of their employees stricken with leukemia.

Excessive exposure, or levels of radiation that are too high, results in acute radiation sickness (ARS). In ARS, which can happen within 24 hours of exposure, radiation destroys cellular structures and DNA. The effects of radiation overwhelm the body's ability to repair itself, leading to compromised immune systems, low blood-cell counts (anemia), and opportunistic infections. Alpha and beta particles are not nearly as dangerous as gamma rays, because they cannot penetrate the skin. One has to ingest an alpha or beta emitter for it to be dangerous, a recent example being the horrific poisoning of Alexander Litvinenko.

Litvinenko was a Russian émigré living in London and a former Russian security service officer. He was admitted to the hospital in November 2006, not knowing the cause of his illness but realizing he had been poisoned, and died three weeks later. Only in the last few hours of his life did investigators discover the causative agent: polonium-210, an intense alpha emitter. They pieced together a story of Litvinenko's drinking from a pot of green tea, laced with the isotope, served at a dinner he had with two Russian officials. Polonium-210 was later found at places those two dinner companions had visited, including in debris in a waste pipe in a hotel room where they had stayed. Despite the scientific evidence, the alleged perpetrators were never brought to justice.

The "invisible phenomenon" discovered by Becquerel powers both good and ill. Radioactivity, like sunlight, is another source of energy in our environment. Carbon-14 is an agent of this energy. One of the first applications of the new, long-lived isotope was to date artifacts found by archeologists. Almost simultaneously, carbon-14 became a tracer, used to define biochemical pathways, notably the fixation of carbon dioxide into organic matter via photosynthesis. The next three chapters discuss these two revolutions in science (both leading to Nobel prizes): carbon dating and the pathways of photosynthesis.

To extend the power of radioactivity from carbon-14, new methods had to be developed, and these new methods came to fruition in the second half of the twentieth century. Scintillation counting and accelerator mass spectrometry, which I consider in chapter 7, allowed researchers to probe biological processes as never before and to extend our understanding of past climates, ancient civilizations, human origins, and a multitude of other applications.

4

DATING

irline pilots and flight attendants, who put in all those hours above 30,000 feet getting us across the country or oceans, are classified as radiation workers (see the previous chapter). The FAA recommends that they limit their flight hours—how much time pilots spend in the cockpit and flight attendants spend serving passengers. As a traveler with your pocket Geiger counter, or with the GeigerCounter app on your smartphone (just kidding), you will notice an increase in tick rate as the plane climbs to cruising altitude. Radiation arriving in the atmosphere from deep space produces those ticks: cosmic radiation, or what are called cosmic rays. Cosmic rays are parts of atomic nuclei such as protons, traveling at nearly the speed of light. The most energetic ones are thought to originate outside our galaxy from exploding stars or black holes. Our sun also sends out cosmic rays.

Most cosmic rays are deflected away by Earth's magnetic field, but they can reach the upper atmosphere near the poles where the Earth's magnetic force lines pass through. Even there, the atmosphere intercepts most of the cosmic rays; the molecules making up our atmosphere absorb or react with them to produce the auroras—the aurora borealis in the northern hemisphere, and the aurora australis in the southern. Without much atmosphere left seven or eight miles up, cosmic rays are much more abundant, and a source of radiation exposure occurring behind the smiles and professionalism shown by flight crews.

Back in the 1930s, when carbon-14 was detectable only as tracks in a cloud chamber, Serge A. Korff, a physicist, began a study of cosmic radiation. How were cosmic rays distributed above the Earth? How abundant were they? At the time, he worked at the Bartol Research Foundation at Swarthmore College, near Philadelphia, not that he spent much time there. To study cosmic radiation, he needed higher altitudes. He traveled the world to various mountaintops, and also launched balloons that could get his instruments even further up, into the stratosphere. The payload in the balloons held Geiger-Müller tubes or other specially designed radiation counters.

When the balloon payloads were recovered after their trip to the upper reaches of the atmosphere and then analyzed, Korff always found an altitude band, about 10 miles up, with evidence of a secondary radiation in the form of neutrons. Korff surmised that these neutrons must have been produced by cosmic rays in their collisions with atoms in the atmosphere. The energy of the cosmic-ray particles was enough to break apart the nuclei in these atoms, producing energetic subatomic particles, notably neutrons. Above the 10-mile altitude band, neutrons escaped into space, because there was not much atmosphere left (or gravitational pull). Below the maximum in neutron abundance, cosmic rays themselves declined, deflected by the Earth's magnetic field. Any neutrons produced by the fewer cosmic-ray collisions at these lower levels in the atmosphere would be quickly absorbed. These two compensating factors—not enough neutrons above, and not many cosmic rays below—left the maximum neutron abundance 10 miles up.

Korff also reasoned that because nitrogen gas was the molecule in the atmosphere in greatest abundance, at about 80 percent of the total, any neutrons set free after a cosmic-ray proton collided with the nucleus of an atmospheric atom would be available to form carbon-14. The sequence, then, has a cosmic-ray proton smashing the nucleus of an atom in the atmosphere and setting free a neutron. The neutron can now enter the nucleus of a nitrogen atom (nitrogen-14) and, freeing a proton, create carbon-14. At the time Korff published his finding, Kamen and Ruben's discovery of the long-lived isotope of carbon was less than a year old. The reaction that Korff surmised was much like what Kamen had worked out with his carboys filled with ammonium nitrate (see chapter 2), with nitrogen gaining a neutron but giving up a proton to make carbon-14.

The carbon-14 atom itself, as we know, does not decay readily, but it also does not hang around independently for long; it is quickly oxidized to radioactive carbon dioxide, $^{14}CO_2$. Cosmic rays act like Lawrence's cyclotron: high-energy particles collide with atoms in the atmosphere, the neutron products of which enter the nucleus of the most abundant molecule in the atmosphere, nitrogen-14, as N_2. Carbon-14 produced this way comes, on average, from only one nitrogen atom in a trillion, but the effect can be measured.

It was not a huge intellectual leap to recognize that if a natural process produces carbon-14 continually in the atmosphere, it would revolutionize science. Willard Libby (figure 4.1), at the University of Chicago, is credited with making the connection in 1946 and with realizing the potential of carbon-14 as a means of dating organic materials, or anything with carbon in it. Libby had been working on these problems before, in the 1930s, but as with many of his cohorts, the war effort interrupted his research.

FIGURE 4.1

Willard Libby in his laboratory.

Photograph courtesy of Richard Longman.

During the war, he worked on the Manhattan Project, part of the over-all program to produce the first nuclear bomb, beginning even before the Japanese bombed Pearl Harbor in 1941 and America entered the war. When the United States dropped the bomb on Hiroshima in August 1945, Libby was finally able to tell his wife what he had been working on.

Carbon-14 atoms produced in the upper atmosphere react with oxygen to become $^{14}CO_2$. This radioactive carbon dioxide behaves like regular (carbon-12) carbon dioxide and is mixed through the atmosphere. Some of the $^{14}CO_2$ finds its way to the troposphere, the atmospheric layer just above the surface of the Earth. There, the radioactive carbon dioxide is taken up by higher plants, trees, and grasses in photosynthesis (see chapters 5 and 6). Most of the radioactive carbon dioxide, however, is mixed into the surface waters of the Earth and is taken up by floating algae, the phytoplankton, in photosynthesis taking place in lakes, lagoons, estuaries, and the ocean itself. Animals eat the plants. We eat the animals, and also the plants. Every living thing on Earth thus becomes radioactive, however slightly. And both plants and animals make hard parts, such as bones, shells, and other calcareous structures, mostly calcium carbonate. Eventually, everything becomes radioactive to the tune of one carbon-14 atom in a trillion carbon-12 molecules (figure 4.2).

To summarize, Serge Korff identified the pathway from the neutrons that emerge as a result of cosmic-ray collisions with atomic nuclei, which then enter nitrogen nuclei and, expelling a proton, create carbon-14. Willard Libby realized that the carbon-14 was created naturally and continuously in the atmosphere. Carbon-14, then, could be used the same way as other radioactive isotopes to date the time of death of living matter: carbon-14 dating.[1] Libby, however, came up against a dilemma. It was one thing for Libby to understand, intellectually, that dating the death of an organism would work. It was quite another to measure the natural abundance of carbon-14 in that dead organism. The methods available in the late 1940s and early 1950s were simply not up to the task. Here's why.

Given the cosmic-ray abundances and the nitrogen atoms available, one can compute that carbon-14 atoms are produced in the upper atmosphere at a rate of two atoms of carbon-14 per square centimeter, every second. That also means that two atoms per square centimeter will decay every second to maintain the balance between production of carbon-14

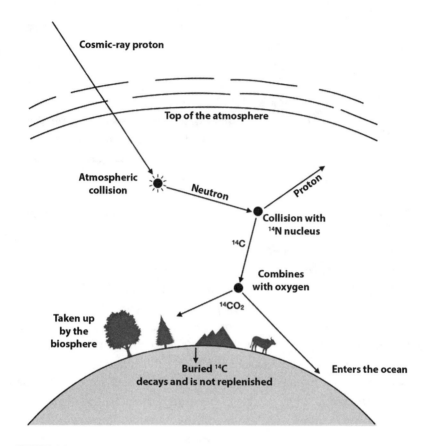

FIGURE 4.2

Production and fate of carbon-14. Neutrons that result from cosmic-ray particles' smashing into atoms in the atmosphere collide with a nitrogen-14 nucleus, expelling a proton and creating carbon-14. The carbon-14 atom is oxidized to $^{14}CO_2$, which reaches the surface of the Earth, enters the ocean, and is assimilated into plants and algae by photosynthesis. From there, it reaches the rest of the biosphere.

and its radioactive decay. For every two atoms produced, two atoms must decay. Given that the ocean covers 71 percent of the Earth's surface, most of the exchangeable carbon at the surface of the Earth is there; it includes dissolved carbon dioxide, bicarbonate, calcium carbonate, and organic carbon molecules in solution in ocean water. (I'll revisit these in more detail in a later chapter.)

Together, the various ocean carbon entities come to about 7.8 grams per square centimeter. The biosphere and detritus on the land part of the Earth plus Earth's atmosphere adds only another 0.7 grams per square centimeter, bringing the total to approximately 8.5 grams of carbon per square centimeter of Earth's surface (Libby 1961). Two atoms per square centimeter divided by 8.5 grams per square centimeter, multiplied by 60 seconds per minute, gives an average radioactivity of 14 disintegrations per minute (dpm) per gram of organic carbon. The expected value for a gram of organic carbon, living or recently dead, would only be about 14 dpm.

On the other side, Libby calculated radiation from the "background," or the environment, to be 500 dpm, about 35 times higher. The amount of carbon-14 in dead material was very low. The background radiation, all around us, was much, much higher. What might seem trivial in radioactive assays today was an enormous problem back then.

In fact, the problem had three parts. First, Libby realized that he needed to concentrate biological or organic material if he was to have any hope of determining its ratio of carbon-14 to carbon-12. Second, as a chemist, Libby realized that, in addition to recently dead organic material, he also needed something old enough that it would be completely free of carbon-14. He could only validate his measurements if he had something that would register as "zero." For his recently produced biological material, which should have measurable carbon-14, he chose methane. Bacteria produce methane, breaking down sewage, the organic matter entering wastewater treatment plants. Then, he devised a means of concentrating that methane chemically. Next, he employed the same methods that he used for the sewage with methane from petroleum. Petroleum methane is so old, millions of years old, that any carbon-14 would have long since decayed away. This was Libby's zero.

For the third part of the problem, he had to figure out a way to get around the presence of background radiation. The background would come from any radioactive substances in the immediate environment, such as from the decay of radium, uranium, or thorium in rocks and soil. Libby calculated that environmental radiation would come to about 400 dpm. The ever-present cosmic radiation from above, at about 100 dpm, made up the remaining background.

For this third problem, Libby did two things, one of them clever and revolutionary. First, he needed to shield his counters. The counters themselves were several Geiger-Müller tubes surrounding the sample chamber, another tube, all attached lengthwise, resembling a bundle (figure 4.3). The central tube, with a sample to be tested, also contained a counter. Then, the counter tubes and sample chamber were encased in thick iron. The central tube was also inside a mercury sleeve for further protection from background radiation. The iron shielded the bundle of tubes from environmental radiation, reducing the total background from 500 to 100 dpm, still not enough to discriminate the natural abundance of carbon-14. Cosmic radiation could still penetrate the shield.

The remaining 100 dpm background from cosmic radiation was still there, so Libby came up with an innovative way to ignore it, to tune it out. The outer counters, the ones surrounding the central, sample counter, would sense both cosmic and sample radiation. The counters were wired

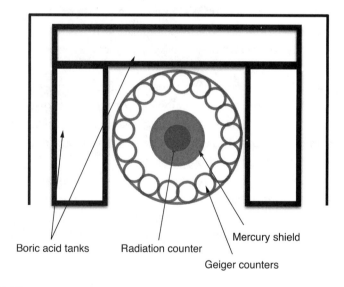

Boric acid tanks   Radiation counter   Mercury shield

Geiger counters

FIGURE 4.3

A schematic of Libby's apparatus showing the Geiger counters arrayed in a circle around the sample, or radiation counter, which is jacketed by a mercury sleeve. Surrounding the counters are other kinds of shielding: boric acid tanks, themselves surrounded by lead and iron casings (see Taylor 2016).

together in such a way that when the outer counters sensed an event, the central counter (with the sample) was turned off for a millisecond. In this way, the cosmic radiation from outside the system could be ignored in favor of the radioactivity of the sample—a clever solution to the problem.

That electronic fix further reduced the background to between 1 and 6 dpm, an acceptable rate. The methane gas from the City of Baltimore sewage gave the expected value of 14 dpm. The petroleum methane registered at zero, as expected. Human waste from sewer lines sent science onward.

Because Earth's magnetic field deflects incoming cosmic rays toward the poles, they are distributed by latitude, with more cosmic rays entering the atmosphere at Earth's magnetic poles. At the poles, cosmic rays reach altitudes where they can interact with the molecules of the upper atmosphere and produce the auroras. If there were a latitudinal variation, it would throw into doubt the dating of any material because the background correction would be wrong.

As a final step, therefore, Libby's team tested whether the 14 dpm value held up despite this strong latitudinal variation in cosmic rays, whose effect should be obliterated by the shorter-term variations in winds and ocean currents compared to the ~8,000-year mean decay rate (and ~5,700-year half-life) of carbon-14. Widely separated samples, geographically distributed from the polar regions to the tropics, showed that the radioactivity was constant, within measurement errors. The supposition that environmental variations would override latitudinal variations in the number of cosmic rays turned out to be correct.

Libby and his students were now ready to test some real samples. When they consulted with archeologists and historians of the ancient world, they were astonished to learn that verifiable dates for artifacts went back only 5,000 years, to the First Dynasty in Egypt. Anthropologists wrote and talked about human habitations, say, from 20,000 years ago, but the archeologists admitted that assigning absolute dates was actually just guesswork. They could only arrive at relative and very approximate dates for earlier artifacts (not that this guesswork prevented archeologists from theorizing about the origins of civilization and human habitations, as I discuss in chapter 8). According to the archeologists, then, the oldest sample for which a date could be historically verified was a date slightly less than the half-life of carbon-14. That verifiable date, at 5,000 years before the present, was fortuitous. Also fortuitous was that the University of Chicago employed an impressive group of Egyptologists.

Libby took bits of wood held at the University of Chicago and in museums from around the world. He used wood found among the brick in the tomb of King Zet and Vizier Hemaka, from ancient Egypt's First Dynasty, as the oldest for which there was a verifiable date. He also took samples from cedar wood used in sarcophagi and wood from the deck of a funeral ship buried with Sesostris III. These were all from early Egyptian civilization. "Younger" samples came from a Persian palace (2625 BCE); from the linen wrapping of the Great Isaiah Scroll, one of the famous Dead Sea scrolls found near there, in Palestine (150 BCE); and from an "overdone" bread roll, cooked as a result of the volcanic eruption that buried Pompeii (79 CE). He also used a sample from a redwood, *Sequoia gigantea*, felled in 1874. The redwood's stump, known as "The Centennial," named for the U.S. Centennial celebration, had about 2,900 annual growth rings, and each could be identified and counted. (I'll examine tree-ring analysis, or dendrochronology, as it is called, in a later chapter.) He also used more recent wood from Douglas fir tree rings. Libby and his coworkers analyzed all of these samples, whose ages had been established independently, using their recently established methods for carbon-14.[2]

Some of their data are plotted in figure 4.4. The radiometric ages do not fall exactly on the theoretical line, but they are close enough to give credence to carbon-14 dating. Uncertainties are on both sides. The theoretical curve is based on a half-life for carbon-14 of 5,568 years, what it was calculated to be at the time, and certainly subject to error. Libby actually thought it should be longer; the half-life today is put at 5,730 years. On the other side, the artifacts themselves may have been inaccurately dated. Each determination produces errors, either from carbon-14 dates or from historical dates.

How does dating these artifacts actually work? We all know, in our current culture, when something—an email, a video, or other social media post—"goes viral." The phrase originates with viral replication and the rapid spread of infection. When an email goes viral, someone relays the message to a follower, and the follower relays the message to his or her several followers, and so on. Or consider a bank account from which you

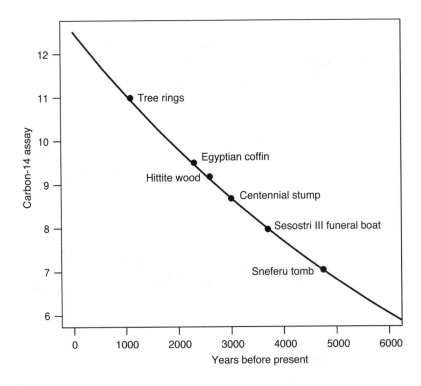

FIGURE 4.4

Examples of carbon-14 dates for archeological samples with known dates (after Libby 1960). The solid line is the expected curve calculated using a carbon-14 half-life of 5,568 years.

withdraw a constant percentage of money each week, or month, or year. If you have $1,000 to begin with, and you withdraw 20 percent after, say, a month, you are left with $800. Another 20 percent withdrawal after another month means $800–$160, leaving $640; another month's withdrawal would leave your account with $512, slightly more than half the original amount after only three months.

Both of these examples show an exponential relationship. The resulting curve of bank funds against months is called a negative exponential. If you have a savings account with compounded interest, your funds increase exponentially, albeit, these days, at a snail's pace. That and the "viral" post show a positive exponential. The constant rate of growth or decay of something is defined, or scaled, by a constant, called e, the base rate of growth for all continually growing or decaying things. The constant e tells us the limit by which something can grow, sort of like a speed limit. Growth or decay is e raised to an exponent representing the rate of increase.

Just like logarithms of the numbers we are all familiar with, based on 10s (0 . . . 10, 100, etc.), there are logarithms based on e, the natural logarithms. Whereas e raised to some amount, e^x, expresses a quantity of x, the natural log of x, $\ln(x)$, is an expression of time. And time is the objective of dating with carbon-14. When I was a student, we had to use tables in the back of our textbooks to look up the natural log of a number, and sometimes interpolate between the values listed. Now, one need only push a button on a digital calculator.

Exponentials permeate everyday life. Human population growth since the beginning of agriculture is a rough exponential increase. Bacterial cells in a petri dish—one dividing into two, the two into four, and so on—multiply exponentially. Exponential growth of bacterial pathogens in the body make them particularly dangerous to one's health, going from a small number to lethal concentrations in a matter of a few days.

The ecosystem of plants, animals, decomposers, and humans takes in carbon-14 through photosynthesis by the plants and releases it through respiration by the same plants, and by animals and decomposers, in a rough balance, or equilibrium. All life takes in carbon-14 formed by cosmic rays from nitrogen-14, and the carbon-14 in living matter decays back to nitrogen-14 at more or less the same rate. Once living things stop metabolizing—that is, they die—the carbon-in/carbon-out process stops, and carbon-14 is no longer taken up. The carbon-14 atoms left in the formerly living matter disintegrate, through radioactive decay, ticking away randomly, decreasing in activity by half in about 5,700 years.

Thus, everything produced from photosynthesis, directly or indirectly, now contains a radioactive clock. If we know how much we have, and the percentage "withdrawal" (decay), we can calculate the length of time since carbon-14 was in balance with the environment—when the flax plant, cowhide, charred wood, manuscript, piece of bone, or other artifact ceased living and exchanging its carbon. Anything organic, or the product of an organic reaction, can be given an absolute date using carbon-14.

To calculate the age for one of these, it works better to use the mean-life, the exponential decay rate of a carbon-14 atom, rather than its half-life. The mean-life for carbon-14 is 8,033 years (the half-life divided by the natural log of 2, or 5568/0.693).[3] The equation is

$$Age = 8033 \times \ln(N_0/N) \text{ Years}$$

In the above relationship, N_0 is the number of carbon-14 atoms at the time of harvesting or death, where Age = 0, and N is number of carbon-14 atoms in the sample today; "ln" refers to the natural logarithm of (N_0/N). We can find N_0, the original number, by assuming that the carbon-14 in the atmosphere today is the same as it was when the organism died, and that the organism was in equilibrium with the atmosphere at the time of its death. The above equation forms the basis of all radiometric dating. It is the same as that used for other radioactive isotopes, including potassium-40 (half-life of 1.2 billion years), uranium-235 (half-life of 700 million years), thorium-230 (half-life of 75,000 years), and others.

Dating a sample based on carbon-14 presents other kinds of problems. As noted previously, the ratio of carbon-14 to the most abundant isotope of carbon, carbon-12, has to be constant over the time period that the organism lived. This assumption is generally easy to accept because the half-life of carbon is hundreds or thousands of times longer than the lifetimes of all but a few organisms. However, as the science of carbon dating matured, it became apparent that there are indeed variations in the carbon-14/carbon-12 ratio in the environment, caused by changes in the rate of production of carbon-14 in the upper atmosphere. These variations in the rate of production of carbon-14, instead of creating a problem with dating artifacts, opened up new avenues of research. For Hans Suess, the variations make up an "interesting parameter."

Suess, an Austrian, began his scientific studies in Germany and Switzerland in the 1930s, researching nuclear power. After the war, he came to the United States, first to the U.S. Geological Survey in Washington, D.C., then to Harold Urey's laboratory at the University of Chicago, and finally, in 1958, to the Scripps Institution of Oceanography, in La Jolla, California. He was one of four who founded the University of California, San Diego (UCSD). At Scripps and UCSD, he established the La Jolla

Radiocarbon Laboratory, devoted solely to the measurement and interpretation of carbon-14 in the environment.

Suess saw three revolutions in the thinking about the use carbon-14 dating.[4] The first was the utility of a constant rate of decay of carbon-14, used to good effect by Willard Libby. The second revolution in thinking came with the realization that geomagnetic and other geophysical phenomena affected the production of carbon-14 in the upper atmosphere, thereby causing natural variations in the abundance of carbon-14. The third revolution used these variations as subjects of scientific research in their own right, to learn more about Earth, the sun, and the solar system. I've told Willard Libby's story of the first revolution. Now I'll tell the stories of the second and third.

■ ■ ■

The neat trick of calculating the age of some organic artifact depends on the rate of formation of carbon-14 in the upper atmosphere being unchanging. The ratio of carbon-14 to carbon-12 has to be constant over the time period of for which carbon-14 dates can be calculated. If it changes, the assay returns the wrong age. Supposing, for example, that a plant or an animal died during a period when less than the expected amount of carbon-14 was being produced in the upper atmosphere. In that case, the plant or animal would appear older than it actually was, with proportionately less carbon-14 left in the artifact or remains. One the other hand, suppose that death occurred during a time when the atmospheric production of carbon-14 was at a maximum. With more carbon-14 being assayed, that artifact would be calculated to be too young.[5]

The rate of formation of carbon-14 does in fact change over thousands of years. Libby was lucky in that the oldest verifiable date from the historians was only 5,000 years, less than one half-life of carbon-14. If he and his coworkers had tested older materials, they probably would have encountered errors that would have cast serious doubt on dating artifacts using carbon-14, and set back the utility of the method by decades. He did mark, however, small anomalies in other carbon-14 dates that he was unable to explain.

Hessel de Vries, a Dutch physicist from the University of Groningen, analyzed tree rings from a variety of locations and figured out that Libby's anomalies could be correlated globally. The anomalies were not random noise, but actual environmental variations that to De Vries had meaning. These variations are now known as "De Vries effects." De Vries, an early researcher in carbon-14 dating, was known for his meticulous methods and commitment. Whereas Libby analyzed his samples as solids, as charcoal, De Vries showed that with proper methodology, carbon dioxide gas would give more precise results. Precision was at a premium when trying to understand the small anomalies in the dates. His untimely death occasioned another tragedy in carbon-14's revolution in science.[6]

The production of carbon-14 molecules depends on the number of cosmic rays entering the upper atmosphere. If that number changes, then the number of carbon-14 atoms produced will change as well. Two factors have to be considered concerning the production of carbon-14 in the upper atmosphere via cosmic radiation: changes to Earth's magnetic field, and variations in the sun's radiation.

Changes to Earth's magnetic field affect the carbon-14/carbon-12 ratio in the atmosphere. Earth's magnetic field arises from the core of the Earth, composed mostly of liquid iron and nickel. Magnetism itself will occur when any two of three things happen, and those two predict the third. Because it is molten liquid, Earth's core has motion, and because it is metallic (mostly iron and nickel), it can also produce an electric current. The electric current (1), in motion (2), produces the magnetic field (3). An alternative example is the everyday electric motor. Here, an electric current (1) is induced by the spinning motion (2) of a magnet (3). For little understood reasons, Earth's magnetic field changes, and sometimes it changes so much that its direction completely flips: magnetic north becomes south, and south becomes north. One of the fortunate things that the magnetic field does is to deflect the solar wind and cosmic rays away from the Earth, shielding us from harmful radiation. If the magnetic field lessens, less solar wind is deflected and more cosmic rays enter the upper atmosphere, generating more carbon-14.

The second factor is that the sun's radiation reaching the upper atmosphere varies. The output of the sun itself changes, and sunspots, occurring and reoccurring at intervals, also affect solar energy. The Maunder Minimum is so named for its discoverer, Edward Maunder, and for the fact that the sun went through a quiet period of lower solar output during the seventeenth century, more specifically during the reign of Louis XIV of France (Eddy 1976). Louis was famously the "Sun King," perhaps making up for the slightly weaker sun during his years on the throne. In any case, a lower solar output translates into a higher galactic cosmic-ray flux, and the creation of more carbon-14 atoms. Sunspots modify the solar wind such that fewer sunspots mean a larger cosmic-ray flux, and more carbon-14 atoms.

There is another source of variability of a different kind, unrelated to the geophysical effects but just as important. Materials with different sources can be mixed within an organism, rock, or other sample. For example, carbon-14 in the deep ocean, the topic of chapter 10, has not been exposed to the atmosphere for hundreds of years. Nevertheless, deeper waters will mix with shallower waters with much younger ages. The mixing of differently aged pools, called the "reservoir effect," can significantly modify age determinations based on carbon-14.

If these natural variations are not enough, we humans, in all our innocence, have altered the carbon-14/carbon-12 ratio. Around 1870, humans started burning fossil fuels for industry, supplying carbon dioxide, a ton of coal at a time, to the atmosphere, and changing it as a consequence. Coal and oil, organic carbon burned for energy, produce carbon dioxide as a waste product, much as we do ourselves in respiration. These fossil fuels are really old, millions of years old. By the time they are extracted from the ground and burned in factories and automobiles, all the carbon-14 has decayed away. The atmosphere today, therefore, has a different isotopic composition than it did a couple of centuries ago, simply by virtue of our adding nonradioactive carbon dioxide to Earth's atmosphere. Absent radioactivity, additions of fossil-fuel carbon dioxide to the environment dilutes the atmosphere with respect to carbon-14 (figure 4.5).

So, fossil fuel pollutes the atmosphere in yet another way. The isotopic dilution of the atmosphere from fossil-fuel burning is called the "Suess effect" after its discoverer, Hans Suess, who, as noted previously, founded and directed the La Jolla Radiocarbon Laboratory at the University of California, San Diego. The dilution of carbon-14 is not as large as might be anticipated because the atmosphere still gets carbon-14 from respiration in the biosphere. Nevertheless, the atmosphere is becoming "older" from human activity. Continued dilution of carbon-14 in the atmosphere will, in the near future, render carbon-14 dating unusable for artifacts

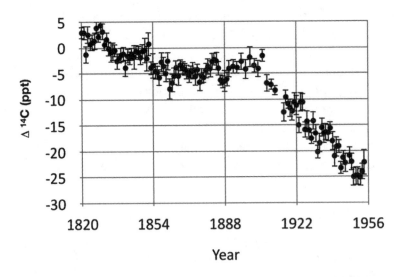

FIGURE 4.5

Radiocarbon content of tree rings of Douglas fir (U.S. west coast), 1820–1954. These data, which depart from expected values, illustrate the "Suess effect," the decline in carbon-14 relative to carbon-12 in parts per thousand ($\Delta^{14}C$, ppt) (after Stuiver and Quay 1981). The values between 1890 and 1915 are two-year averages.

younger than several hundred years. The atmospheric age at the end of this century will be 2,000 carbon-14 years old! On the flip side, carbon-14 determinations have helped identify the cause of air pollution in many cities. It cannot be argued that a weather phenomenon or trees are producing urban smog if the aerosols have no carbon-14.

We have also increased the carbon-14 on Earth in the last 50–60 years. Atomic and hydrogen bomb testing in the 1950s and early 1960s created a spike in carbon-14 atoms in the atmosphere. Nuclear bombs, both fission and fusion devices, release neutrons, which act in the same way as cosmic rays: they collide with nitrogen-14 atoms to produce carbon-14.[7] Above-ground atomic bomb testing spiked the concentration of carbon-14 in the atmosphere (relative to carbon-12), increasing first to 200 parts per thousand (ppt) in $\Delta^{14}C$, and then to 1,000 ppt. The cessation of testing in the mid-1960s meant a slow decay in the atmosphere as carbon-14 mixed in the environment, or was taken up in other processes. Most

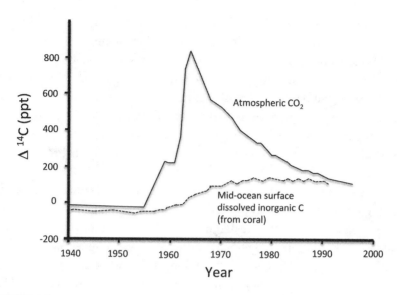

FIGURE 4.6

How atomic bomb testing affected the carbon-14 content of the atmosphere and its entry into the surface ocean (measured in coral skeletons in the southwest Pacific Ocean). The unit is the amount of ^{14}C relative to the amount of ^{12}C in the atmosphere (standardized), in parts per thousand ($\Delta^{14}C$, ppt) (after Druffel 2016).

important, "bomb" carbon-14 mixed into the ocean (lagged by a decade or so) (figure 4.6) and was taken up by photosynthesis. On the timescale of Earth processes, the essentially momentary enrichment of carbon-14 also led to new science, particularly in oceanography, as we'll see in a later chapter. In the parlance of tracer biology, bomb testing lasting only 10 years or so was a "pulse-chase" experiment: momentary exposure to a tracer isotope, followed by where it goes in the system and how long it takes after the exposure ends for the system to come to an equilibrium. When dating an artifact, the carbon-14 assays are sometimes called "prebomb." Carbon-14 ages are usually referenced to 1950, just before nuclear tests in the atmosphere began.

Changes to solar output, Earth's magnetic field, and reservoir effects obviously complicate carbon-14 dating. But the complications have had two good outcomes. First, as De Vries found, carbon-14 variations are correlated over wide areas, signaling a common source. The variations seen in any sample could be used as a means to study solar and geo-physical variations as scientific fields of endeavor on their own. Past variations in the production of carbon-14 mark changes to solar output, with corrections applied for geomagnetic changes. Changes in solar output can be used in studies of climate change in Earth's past. An advantage to these variations is that they indicate global phenomena, not something local such as might be found with a tree. The geophysical studies involving carbon-14 represent Hans Suess's third revolution in carbon-14 dating.

Another outcome of the variations in carbon-14 is that instead of relying on theory, exponential radioactive decay, as the means to date materials, scientists have created a calibration curve for carbon-14. By "calibrate," I mean the scientific definition of "changing a relative value into an absolute." "Calibrate" means taking a carbon-14 date and putting it on a calendar. Independent means exist to calibrate the carbon-14/carbon-12 ratio back several thousand years. Remember that Libby himself used the stump of a huge redwood, dating its lifetime by the annual growth rings, about 2,900 of them. The giant sequoia, however, is a mere baby compared to the bristlecone pine tree, called the "world's oldest living thing." One tree had annual growth rings going back 4,000 years. You count the rings, measure the carbon-14 in the wood of each ring, and construct a calibration curve. Further back than 3,000 years, the measurements deviate from the expected theoretical curve, and this of course has led to research into why this is so. The calibration curves serve both to correct dates based on carbon-14 and as a way to study the geophysical causes of natural variations in carbon-14 itself. Here I review, briefly, a few of the ways the calibration curve for carbon-14 is constructed. I'll revisit these as climate indicators in chapter 13, where I discuss carbon-14 and climate.

Minze Stuiver is a geochemist (now retired) at the University of Washington (UW) who, over his career, worked closely with Hans Suess. Stuiver built the Quaternary Isotope Laboratory at UW, the working part of which is a lead-lined room 30 feet below ground—as certain an escape from environmental and cosmic radiation (and an escape from just about everything else) as there ever could be. Stuiver finds variations in carbon-14 over the past millennium that correlate with variations in sunspots, such as the Sporer (1500 CE) and Maunder minima (1700 CE), after he has corrected for any geomagnetic variations. As expected, the two minima show maxima in carbon-14 production during those times. The tree-ring record reveals the history of solar output. From the 1,000-year record, it is then possible to extend back in time, for as long as the tree-ring records are available, the variations in solar output and their possible effects on Earth's climate.

Tree rings are accurate to the year, but they only go back so far. Carbon-14 dates can be independently verified from other sources. Sediment "varves" in lakes are distinct layers of sediment, laid down each year, much like tree rings. In Lake Suigetsu, a deep lake in Japan with anoxic bottom waters, the varves can supply more or less accurate dates from 25,000 to 30,000 years before the present. Being anoxic, the layers are completely undisturbed; there are no organisms moving about to churn up the bottom sediments. Corals can also be dated. Corals produce growth rings much as trees do, and for validation, corals can be dated using uranium isotopes, which do not have the kinds of isotope-production variations that carbon-14 has. Similarly, limestone deposits in caves, called speleothems, are dated with uranium isotopes and correlated with carbon-14.

Beryllium-10 is yet another isotope that is used to calibrate carbon-14 dates. Beryllium-10 is produced from collisions with nitrogen and oxygen atoms in the upper atmosphere. Beryllium-10 both combines quickly with aerosols and has a long half-life. Combining with aerosols means that beryllium-10 will quickly find its way down to the Earth's surface. These two attributes, a short life in the atmosphere followed by a long half-life in the environment, make beryllium-10 an ideal isotope for calibrating carbon-14 variations.

The calibration curve resulting from all these—tree rings, corals, speleothems, sediment varves—is something of a conglomeration. The curve for carbon-14 dates, like most anything else, has uncertainties and errors. Calibrating carbon-14 dates is a community effort, producing the curve "IntCalXX" where XX stands for the last two digits of the year in which it is published, the most recent being IntCal13. The resulting dates are fuzzy, plus-or-minus several or more years, because of measurement error or uncertainty in the way in which the calibration takes place. Carbon-14 dating has become a science unto itself.

■ ■ ■

Libby's inception of carbon-14 dating in the early 1950s spread "like wildfire" (Willis 1996) to virtually all archeologists and anthropologists. Dating artifacts by carbon-14 removed speculation, firmed up otherwise fuzzy notions, allowed an actual calendar of events, and also trashed many ideas held dear by archeologists, climate scientists, and anthropologists. Dates have been recorded for any number of archeological, geological, and biological materials and artifacts. Carbon-14 became not only a precise indicator of the dates of artifacts, but also a tool for understanding variations in solar output and, by extension, Earth's climate.

Many further applications of carbon-14 dating occurred later, in the 1970s and 1980s, after advances in assaying were made, in scintillation counting and accelerator mass spectrometry, the topics of chapter 7. In chapter 8, I consider two of the more famous applications of carbon-14 dating, the Shroud of Turin and Ötzie the "Ice Man," as well as some others. As with dating itself, archeology went through a parallel set of revolutions in thinking about human origins and civilization. Carbon-14 was the basis for those transformations.

However, the next turn of the scientific wheel with carbon-14 came in biochemistry, the first steps in the fixation of carbon dioxide into organic matter in photosynthesis. Like the development of carbon-14 dating, this revolution happened in the late 1940s and early 1950s. Why is photosynthesis, the creation of organic matter from carbon dioxide and water, important? How did carbon-14 revolutionize the study of photosynthesis and, by extension, all biochemical pathways?

Photosynthesis, which evolved a few billion years ago, fundamentally changed the Earth and allowed pretty much everything we see around us today. Natural philosophers and scientists since the time of the Greeks have probed its workings, and by the late eighteenth century had worked out the basic equation of ingredients and products. But then they hit a wall. The actual reactions happening in photosynthesis had to await the discovery of carbon-14.

5

PHOTOSYNTHESIS

Geologists put the origin of the Earth, when Earth was "born," at approximately 4.6 billion years ago, about the same time as the creation of the solar system. That initial period in Earth's history is called the Hadean Eon, after the god of the underworld, Hades. Experiencing the Hadean was probably like our version of hell, or maybe Dante's. Earth, recently accreted from smaller entities, sizzled with a molten surface. Comets bombarded the Earth as rock masses and water churned. Over time, the new planet cooled and began to differentiate into major layers: a surface crust, a mantle surrounding a hot, molten outer core, and a solid inner core. On geological timescales, cooling happened quickly. Geochemical evidence indicates that very early on, Earth had an ocean. Carbon dioxide, ammonia, and water vapor made up the early atmosphere, and Earth had a fairly strong greenhouse effect, moderating temperatures. At one point during the Hadean, another "protoplanet" in the nascent solar system collided with Earth, knocking it off its vertical axis, setting it spinning, and creating the Earth-moon system.[1]

The Hadean Eon came to an end with the beginnings of life, about 3.8–4 billion years ago, only about 500–700 million years after the formation of the Earth itself. The first evidence of the use of light energy to fuel life, a form of photosynthesis, is found another 500 million years after that, beginning the Archean Eon. The early aquatic organisms put the primitive-Earth conditions to their own use, using available sulfur compounds to generate energy and relying on a primitive pigment system for capturing the light.

A billion years after Earth's formation, arguably the biggest and most significant event occurred in Earth's history: the invention of a kind of photosynthesis that produced oxygen. Combining carbon dioxide and water in the presence of light changed everything. Photosynthesis not only synthesized organic matter but, in the process, it also expelled oxygen into the environment. "Oxygenic," or oxygen-producing, photosynthesis was fueled by the sun, creating "fixed" carbon—organic matter.

The two molecules used in photosynthesis, carbon dioxide (CO_2) and water (H_2O), are interesting in themselves. In the water molecule, we have one of the most reactive elements in the universe, oxygen, in combination with the most abundant in the universe, hydrogen. In CO_2, as I discussed in chapter 2, we have carbon, the element most conducive to matter's variety and life's complexity.

If you look, there is very little in the environment around you that is not, either directly or indirectly, a product of photosynthesis. Plants, of course, photosynthesize for their own growth and metabolism. But animals, humans, and indeed all of civilization come from "solar power." Cell phones, fossils, brick buildings, books, *Hello, Dolly!*, the pyramids, Hoover Dam, brewed coffee, drugs—none of these could have happened without the evolution of photosynthesis. Oxygenic photosynthesis set Earth on an entirely new trajectory, unique in the solar system, if not the whole universe, a trajectory for the evolution of life.

Oxygen also carried the mechanism to maintain and perpetuate that life. Oxygen (O_2), as a diatomic molecule, reacts with the sun's incoming ultraviolet (UV) radiation, making ozone (O_3). Ozone also preferentially absorbs light at UV wavelengths, breaking down and, in the process, protecting life from UV light's harmful effects. Photosynthesis, while oxygenating the atmosphere on the early Earth, also removed carbon dioxide, sequestering the carbon as organic matter, reducing the Earth's greenhouse effect, and helping to create Earth's milder—that is, less hot—temperatures.

Given the complexity of the process, especially the creation of chemical-bond energy from solar energy, it can seem surprising that photosynthesis happened at all. But organisms had a few hundred million years to figure it out, and to build on simpler schemes. Bacteria were around very early, and bacteria have learned to do pretty much anything. We do not know how life—which is to say, the first single-celled self-replicating organism—arose from nonliving matter.[2] But once there, those early life forms came up with a variety of ways of interacting with their environment to make energy and propagate.

Biogeochemists have pieced together a sequence of ecosystems, all designed for early organisms to extract energy from molecules in their environment. Like the reverse processes of photosynthesis and respiration, these are reduction and oxidation reactions. Reduction frees electrons to create chemical energy and organic matter. Oxidation reactions release that chemical energy from organic matter.

About 4 or 4.2 billion years ago, autotrophic, or "self-feeding," bacteria and archaea gained energy through chemosynthesis. Archaea used molecular hydrogen (H_2) coming out of hydrothermal vents at mid-ocean ridges, and combined it with CO_2 to produce organic matter and evolve methane. Bacteria in those days, instead of hydrogen, used metals to do pretty much the same thing. No oxygen was involved, and no light. Other bacteria, bacterial heterotrophs, organisms that require food to be "made for them," extracted energy from the organic matter produced, and evolved methane, an early form of anaerobic respiration.

Later, when the autotrophs moved up into surface waters or onto land, they evolved pigments and the capability to use light. They used solar energy with hydrogen sulfide (H_2S) to reduce CO_2 to produce organic matter, with sulfate as a waste product. Similarly, some other organisms used iron ions (Fe^{2+}, ferrous iron) and light to reduce CO_2, to make organic matter, with ferric hydroxide as a waste product. For the reverse oxidation process, still other heterotrophs did another kind of anaerobic respiration, essentially reversing these reactions, and extracting energy.

The early ecosystems, the age of bacteria and archaea, go back almost to the origin of Earth itself.[3] Even so, we have not escaped it, or evolved from it since then, except for the thin biological veneer of plants and animals (and the even thinner veneer of civilization). We remain ruled by bacteria, from those in our gut to those that remake the things we use to live and to keep the world spinning its cycles. Archaea are no less diversified. They are functionally similar to bacteria, but genetically as different from them as we are.

These forms of chemo- and anoxygenic photosynthesis went on for a good bit of time, a billion and a half years. Somewhere along the evolutionary path, but long after chemosynthesis and anaerobic respiration, "only" about 2.3 billion years ago, some bacteria evolved a new set of pigments, probably to catalyze further redox reactions to produce the chemical energy needed for growth and reproduction. Here, we have the culmination of sorts. Instead of using the rarer hydrogen sulfide, which is only produced in specific places (hydrothermal vents and springs), these bacterial forms evolved to use water (H_2O), which was everywhere—a definite competitive advantage. Sunlight was also ubiquitous, provided the water was shallow enough for the light to penetrate. The big change was that in using water, the pigmented bacteria evolved oxygen as a waste product—oxygenic photosynthesis.

We call these pigmented bacteria "cyanobacteria," for their cyan, or blue-green, pigment. The old name for these organisms, still in use, refers to them as "blue-green algae," although, taxonomically, they are not algae at all, and not even in the same domain of life. The cyanobacteria evolved the capability of using electrons from water for energy, in the form of hydrogen ions, and using this reducing energy to create energy-rich molecules, ultimately to fix carbon—in short, oxygenic photosynthesis (figure 5.1). The reverse, the oxidation step (respiration), never takes place. The oxygen produced by these early photosynthesizers becomes a waste product and is thrown off. Indeed, oxygen, being so reactive, can become poisonous if not expelled from the cells.

Fossils from those times, 2.3 billion years ago, are almost nonexistent. Instead, we estimate the ages indirectly, from the geological occurrence of oxygen. Because oxygen is so reactive, the evidence comes from oxygen combining with ferrous iron, a major element in Earth's crust, to produce and precipitate iron oxides. Iron oxides are found in rocks that have ages indicating the beginnings of photosynthesis.

That period of oxygen output to the ocean and atmosphere is called the "oxygen catastrophe." Evolution and change always deal out winners and losers; in this case, the "catastrophe" caused a major extinction of those bacteria and archea that made their living without oxygen, called anaerobes, for which oxygen was poison. Anaerobes have not disappeared, but all that oxygen meant they no longer dominated life on Earth.

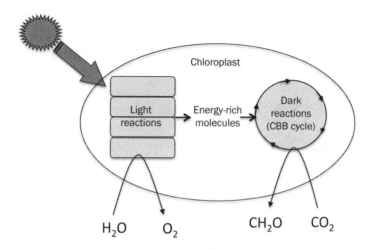

FIGURE 5.1

The overall scheme of photosynthesis. Water (H_2O) is split in the presence of light—the sun—producing free oxygen (O_2) and freeing energetic hydrogen ions ("light reactions") in pigment layers. The hydrogen ions are used to create chemical energy in the form of energy-rich molecules, such as adenosine triphosphate (ATP) and nicotinamide-adenine dinucleotide phosphate (NADPH). These molecules are used in the energy-requiring reactions of the Calvin-Benson-Bassham (CBB) cycle to fix carbon dioxide (CO_2) in transformations not requiring light (the "dark reactions"), and produce carbohydrate (CH_2O), or sugars. (Early photosynthesizers did not have intracellular organelles such as chloroplasts.) The CBB cycle is the topic of the next chapter.

If we compress Earth history to the period of a 24-hour day, the evolution and succession of the early ecosystems happened in only the first four hours or so. Cyanobacteria ruled the microbial world for billions of years after the oxygen catastrophe, and at least in the ocean, they have not relinquished that role. Other groups that photosynthesize—notably the diatoms and dinoflagellates, not to mention land plants—appeared much, much later, 300–400 million years ago, around the time of the dinosaurs. Some of the more ancient forms of cyanobacteria, *Synechococcus* and *Prochlorococcus*, tiny and single-celled, dominate the community of photosynthesizers in the ocean today. Strangely, *Prochlorococcus* (from the Greek, meaning "small green berry") was identified in ocean samples only about 30 years ago.

The modern story of photosynthesis begins about 400 years ago. A common belief before that time, dating to the Greek philosophers, was that plants depended on soil and roots for growth. The roots were primary; the leaves were only there to shade the roots.

■ ■ ■

In the early part of the seventeenth century, as the Thirty Years War raged over central Europe, Jan Baptist Van Helmont, at his home near Brussels, weighed a quantity of soil and humus and put it in a bucket. He then weighed a small willow tree and planted it in the bucket with the soil. He carefully measured and added water to his potted plant, and he continued to add a measured quantity of water (rainwater or distilled water) daily, for the next five years. At the end of this time, he weighed the tree and the soil. The soil weighed just as much as when he started five years earlier, but the willow increased in weight by 164 pounds, leading Van Helmont to conclude that the willow's bark, roots, and leaves came only from the water.

Van Helmont got it half right. Like many scientists of his day, he was independently wealthy, having married into a noble Belgian family, which allowed him to pursue his interests in the natural world. He is sometimes called the father of atmospheric chemistry, even claiming origin of the word "gas." Ironically perhaps, he recognized that the gas given off by forest soil or burning charcoal was dangerous to one's health. We now know that gas to be carbon dioxide, the result of respiration and oxidation of organic matter, a process the reverse of photosynthesis.

Years later, in Bologna, Italy, at the other end of Europe, another early scientist, Marcello Malpighi, a microscopist, discovered that insects had no lungs. They exchanged gases with the atmosphere by way of tubes that opened at the surface of their exoskeleton. These are now called "Malphigian tubules" in honor of his anatomical discovery. When he placed some leaves underwater, in the light, he observed similar structures, pores. He also observed bubbles forming on the surface of the leaves and surmised that plants, like insects, had structures to exchange gases with the atmosphere. The pores are what are now called stomata.

For photosynthesis, that's pretty much where things stood until the mid- to late eighteenth century, during what can now be regarded as a flurry of experimentation. Stephen Hales, in England, like Malpighi, showed how plants could interact with the atmosphere. He figured that plants must get their "nourishment from air" and also speculated about the importance of light. Then Antoine Lavoisier discovered oxygen (as important to combustion), and Joseph Priestley showed in his famous bell jar experiments that "dephlogisticated air" (later identified as oxygen) came from plants, finding that plants both make and absorb gases.

Jan Ingenhousz (1730–1799) put these various observational threads together and is credited with actually discovering photosynthesis in the late 1700s. Ingenhousz was born in the Netherlands, in Breda. Being Catholic in a Protestant country, he attended university in neighboring Belgium, at Louvain, becoming a doctor of medicine. He did, however, spend a final two years at the University of Leiden, where he developed an interest in electricity. On an invitation from Sir John Pringle, the royal physician, he left the Netherlands for England, where he befriended Priestley, as well as the visiting Benjamin Franklin, and Henry Cavendish, an early atmospheric chemist. From Franklin he learned about "variolation," a technique for immunizing people against smallpox (variola) by inoculating them with fluids from people with mild symptoms of the disease.[4] Variolation was controversial, but Ingenhousz eventually became prominent in Europe for his successes with the method in England. He was invited by the Habsburg empress Maria Theresa, herself scarred by smallpox, to inoculate her family, even though doctors in the Austrian Empire were strongly opposed to the technique. Ingenhousz stayed in Vienna as court physician for 10 years and, for his service to Maria Theresa's family, received a healthy annual stipend, making him an independent man of means.

It was not until the 1770s, late in his life, that Ingenhousz returned to England, to Joseph Priestley's lab, Bowood House, in Calne. He repeated Priestley's experiment, in which a plant could maintain a burning candle in the light but extinguish it in the dark. Like Malpighi, he observed bubbles forming on green leaves held underwater in the light, and then noticed that the bubbles stopped forming after a while in the dark (figure 5.2).[5] Because he worked with Priestley, he subscribed to the theory of phlogiston and "dephlogisticated air." After visiting Lavoisier in Paris, however, he came to the conclusion, in the late 1780s, that it was oxygen that was

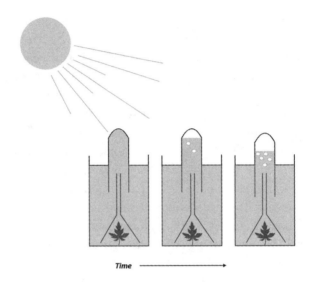

Time ———————→

FIGURE 5.2

Schematic representation of Jan Ingenhousz's experiment, a basic eudiometer, a device used for measuring the volume of gases, and an early version of manometry. In this version, shown as a sequence from left to right, light illuminating a submerged plant causes gas bubbles to displace the liquid in the tube held above.

in the bubbles coming from his plants, and later still, that the gas released by the plants in the dark was carbon dioxide. He further noted that more oxygen was released during the day than carbon dioxide produced at night, and so concluded that the plants were getting their carbon from the air, solving Van Helmont's riddle.

Ingenhousz's results were painstakingly arrived at, only after much careful experimentation. Thus, he represents a more modern, pragmatic, and advanced type of scientific thinking than Priestley, who was a clergyman and a natural philosopher in the traditional sense. Strangely, Priestley never acknowledged his collaborator's contributions and stuck to his phlogiston theory even when it was found to be untenable; he carried those ideas to his death. Another contemporary, Jean Senebier, a pastor in Geneva, completed experiments similar to Ingenhousz's, but published a few years after Ingenhousz's seminal work. Some think he stole Ingenhousz's ideas. Senebier, being a religious man, could not conceive of plants giving the world "bad air" (carbon dioxide) because, in his mind, it went against God's plan for nature. Furthermore, Ingenhousz was probably the first to recognize the importance of photosynthesis in supporting animal life and pretty much everything else.

This abbreviated history of research summarizes the discoveries surrounding how plants feed themselves to grow. It is interesting that it began in the seventeenth century, when observation began to replace textual authority, the time of Galileo's observations through his telescope and Kepler's observations of the orbit of Mars. Europe in the seventeenth century experienced tremendous upheaval during the religious wars, which required political and governmental organization and contributions by scientists. In a somewhat wrongheaded way, it seems, war aided scientific advance.

By 1800, thanks to Van Helmont, Malpighi, Ingenhousz, and the rest, the essential ingredients of plant photosynthesis were known—at least that it was associated with water, carbon dioxide, sunlight, and the green parts of plants, all of which were needed to yield life-giving oxygen. The importance of water was contributed by Van Helmont; Ingenhousz demonstrated the necessity for carbon dioxide, the evolution of oxygen, and the influence of light. It was not until the second half of the nineteenth century that carbohydrate, the last entity in the equation, was determined to be the first organic product of photosynthesis. The significance of carbohydrate came about both indirectly, from trying to balance the exchange of carbon dioxide and oxygen, and directly, from observing the formation of starch crystals in leaves held in the light.

The overall chemical equation for photosynthesis, then, has carbon dioxide plus water making carbohydrate and molecular oxygen, in the presence of light. The chemical equation could be balanced for the inputs, carbon dioxide and water, and the products, oxygen and carbohydrate. About what happened in between, nothing was known. Where did the oxygen come from? Was it from the water molecule? Or from carbon dioxide? And most important, how was the carbon in carbon dioxide converted to carbohydrate? Eugene Rabinowitch, a twentieth-century researcher in photosynthesis, likened the situation to an automotive engine perceived by those of us who are not mechanics (Rabinowitch 1948). We know that gasoline goes in and exhaust gases come out. The automobile's dashboard indicates the behavior of some of the engine's variables, such as temperature and rpm. Analogously, for photosynthesis, we can monitor things like temperature, light intensity, and the supply of nutrients. But what goes on under the hood? How does the "engine" work?

The behavior of photosynthesis under these external factors could be studied, and provided a few clues with physiological meaning—for example, the difference between processes in the light and dark, and how many short flashes of light were required to produce a molecule of oxygen. An early isotopic experiment produced an important clue. When water and carbon dioxide (actually, its chemical equivalent, bicarbonate) were both labeled with oxygen-18 to different levels, the evolved oxygen came from the water, and not from the CO_2. This was one of Martin Kamen and Sam Ruben's early experiments in the late 1930s. But for about 100 years, from the mid-nineteenth to the mid-twentieth century, researchers only knew, as with that automobile engine, what went in and what came out—little of the actual mechanism. The discovery of carbon-14 as a long-lived isotope of carbon changed all that, and began a revolution in science.

6

CALVIN'S CYCLE

he lollypop. "Spots on paper." Microalgae (*Chlorella*). Carbon-14. At
least we recognize the last of these crucial ingredients in the quest
for the pathway of carbon in photosynthesis.

The lollypop? The experiments with leaves were not working, and
Andrew Benson needed some other device for figuring out how plants
took in carbon dioxide from the air and turned it into sugar. Benson, a
biochemist, was one of two assistants to Melvin Calvin, and had been at
the Berkeley Radiation Lab since before the war. The other was James
Bassham, who joined the group as a graduate student after spending the
war years as a U.S. naval officer. With Ruben having died a few years
earlier and Kamen dismissed (see chapter 2), Lawrence put Melvin
Calvin (figure 6.1) in charge of photosynthesis research in 1946, and
Calvin created the Bio-Organic Group at the Radiation Laboratory. The
lollypop, Benson's contribution to the effort, was a glass chamber, looking
very much like a giant lollypop, containing the experimental subjects for
photosynthesis research; as we shall see, it played a key part in figuring
out the fixation of carbon into organic matter.

Up to that time, isotope experiments to discover the path of carbon
in photosynthesis had used oxygen-18, a stable isotope of oxygen, and
carbon-11, radioactive, but with a half-life of only 22 minutes. Even so,
important progress had been made. Labeling both carbon dioxide (CO_2)
and water (H_2O) with oxygen-18, Kamen and Ruben had found that the
oxygen evolved in photosynthesis came from the water molecule and not
from carbon dioxide.

FIGURE 6.1

Melvin Calvin in the laboratory. (Wikimedia Commons)

Any useful results using carbon-11 are a tribute to Kamen and Ruben's planning, experimental expertise, and organization. When the cyclotron produced some carbon-11, Kamen would call Ruben, waiting down the hill in the Rat House, to say he was on his way. Of course, the cyclotron being what it was—temperamental—Kamen's call might be "Cyc's sick, Sam" to say there was no isotope coming after all (Morton 2008). But when he was able to produce some, Kamen would hurry down to the Rat House, a few hundred feet away, and hand Ruben the isotope through the doorway. (Kamen, as noted previously, was not allowed inside because he was radioactive himself, hence a potential source of contamination.) Ruben, for his part, would have everything ready for the treatment and analysis—the solvents, pipettes, absorbent paper, reagents, and the counter recording background radiation. Kamen (1986, 86) described the scene:

> Anyone looking in on the Rat House when an experiment was in progress
> would have had the impression of three madmen hopping about in an

insane asylum, what with the frenzied activity punctuated by loud classical music from the radio monitor [to check for electrical disturbances],
and Sam's yells to get on with it and hand him samples while he sat at
the counter table.

They had only an hour or so before they would lose the isotope's signal,
its radiation decayed away. Added to the frenetic activity was the after-
hours bookwork of devising and looking up analytical methods for the
compounds they hoped to identify.

Even with the time handicaps, Kamen and Ruben's experiments established that carbohydrates were formed in the light, and for a short time
thereafter in the dark. But for longer periods in the dark—hours—no
carbohydrates were produced. They used a molecular-weight approach
to figure out if the carbohydrates, intermediates in the photosynthetic
pathway, occurred in any kind of sequence, and to analyze for the
weights, they needed an ultracentrifuge. Stanford University, across San
Francisco Bay from Berkeley, housed the nearest one. They were looking at a 50-minute drive against a 20-minute half-life for their products
of photosynthesis. Maybe the police could escort their car and samples?
Maybe carrier pigeons? Instead of these extreme measures, they finally
found a comparable machine only 10 minutes from the Radiation Lab, at
a Shell Oil facility in Emeryville. But even this short distance proved to
be unworkable. The short half-life of carbon-11 aggravated the problem of
trying to separate any of the metabolites resulting from their experiments.

Some of Calvin's group's first experiments into the carbon-fixation
pathways of photosynthesis used leaves, probably an obvious choice. A
gaze out any window will show you leaves, stems, and branches of plants,
swaying with the wind, and in the sun, silently and continually, breath-
ing in carbon dioxide from the atmosphere, synthesizing everything the
plant needs to grow, while expelling the extra oxygen. Of course, having branches, stems, roots, and flowers, in addition to leaves, complicates
matters. And some initial experiments proved the point. Photosynthesis
happened too quickly for intermediate compounds to be discerned by the
time a leaf could be analyzed.

So Calvin's group, like Kamen and Ruben before them, resorted to
microalgae. The tiny organisms, within their single cells, do pretty much

everything the "higher" plants do, and sometimes more. Microalgae are no longer classified as plants, but are included in the kingdom Protista, grouped with *Amoeba*, *Paramecium*, and other characters from high school biology labs.[1] Protista includes all single-celled organisms that have internal structures: organelles, like mitochondria, the engines of respiration; the nucleus, where genetic material is stored; and, in microalgae, chloroplasts, the organelle where photosynthesis happens.

The principal model organism Calvin's group used was *Chlorella pyrenoidosa*, a common chlorophyte, or green alga. No, "green alga" is not redundant. Most algae are not green, like typical land plants, but red, golden-yellow, orange, or, as mentioned previously, for bacteria, blue-green. Botanists classify algae according to color, and the color comes from a dominant pigment. For *Chlorella*, chlorophyll-b gives the cells a "grassy-green" color. The most important pigment, the principal pigment in photosynthesis, is chlorophyll-a, which is blue-green, but it is generally masked by other chlorophylls, or by what are called "accessory" pigments. Accessory pigments help the photosynthetic process in various ways, such as extending the colors of light that the alga can use in photosynthesis, or protecting the cell from ultraviolet radiation or light that is too intense. Accessory pigments include the different varieties of chlorophyll (e.g., chlorophyll-b, -c, or -d), and also xanthophylls and carotenes, and come in a variety of colors. Accessory pigments become prominent and exposed in land plants in autumn months, as leaves go dormant and the chlorophylls breakdown, leading to the autumn color displays in temperate forests.

The disadvantage in using *Chlorella* was that you could not get your hands on it. Microalgal cultures existed as something looking like a green solution. For Benson, however, these little photosynthesis powerhouses were ideal. They could be easily cultured in the laboratory, and they grew fast, producing useful population numbers in a matter of days, unlike Van Helmont's years for his willow tree. Using a method borrowed from culturing bacteria—called, appropriately, "continuous culture"—Calvin's team could easily keep a ready supply of *Chlorella*. Further, there were no separate organs for photosynthesis and for nonphotosynthetic functions. And because they were analyzing millions of cells at a time, any variation in individual cells would be averaged out.

Using carbon-14 was, of course, a given. But they added another critical item to their analytical arsenal. In 1943, a laboratory at the University of Rochester developed procedures for separating various components of organic chemicals. The researchers' interests were amino acids, the component molecules of proteins. A student at Rochester, William Stepka, brought the technique with him to Berkeley when he became a graduate student with the Bio-Organic Group. Calvin and Benson immediately saw the potential and applied essentially the same methods for carbohydrates, and in combination with carbon-14.

The separation method the Rochester researchers developed is paper chromatography. Chromatography refers to a separation of colored compounds. An absorbent paper is spotted with an organic substance, allowed to dry, and then placed in a bath of solvent, usually alcohol, so that only the bottom strip of the paper is wetted. In complex mixtures of organic materials, each molecule will migrate with the solvent up the paper, but only so far. How well the molecule dissolves in the solvent governs its travel up the paper. It might remain near the original spot or travel some distance. The different molecules in the mixture will separate, and each spot can be cut out, drained off the paper—"eluted"—and analyzed further for its properties and identification. Back in the day, salesmen of chromatographic systems used flavored drink mixes to demonstrate separations, pouring them through a column, or spotting them on paper. The colors, otherwise unseen, separate, and it is fun to watch. It will also put you off powdered drink mixes forever.

The method can be taken a step further for complex mixtures. Two British scientists, A. J. P. Martin and R. L. M. Synge, added another feature to the process, calling it two-dimensional paper chromatography.[2] The paper, developed with its individual spots in one direction, is then rotated on its side (90 degrees) and put in another bath, with a different solvent, and the process is repeated.

Most biochemists at the time Calvin and his team did their experiments looked down on paper chromatography as nothing more than "spots on paper." For the Berkeley team, it contributed to their success in a big way, along with the lollypop and microalgae. Paper chromatography's usefulness has not faded (sorry) with the years. It is still used today for organic chemical separations and assays.

You can probably guess that chemical analysis by paper chromatography was long, tedious, time-consuming, and subject to error and procedural dead ends. A process that took a couple of days of waiting for compounds to migrate and separate, often in two dimensions, now takes only a few minutes, with "high-performance liquid chromatography" or HPLC. In the method used in the 1940s, large half-meter-square sheets of absorbent paper were washed and then spotted with some extract. The chromatograms were developed in tanks for 12 hours or so and then dried. After drying, they were rotated 90 degrees and placed in different troughs with different solvents for another 6 to 10 hours. If the spots overlapped, they had to be rechromatographed using different solvents.

Benson was the team member who developed the solvents needed to separate out the compounds. He regularly carried a sharpened pocket knife, always ready to cut spots from a sheet. Like many organic solvents, some of the ones they used smelled pretty bad. The solvents infused their clothing even under their lab coats. The lab members got used to them, but not anyone else. On one occasion, the group was asked to leave a Berkeley movie theater after complaints from the other moviegoers.

Bassham was the specialist in carbohydrate chemistry. It was his job to identify the paper chromatography spots—not an easy task. Candidate compounds had to be separated using identical chromatographic methods, producing a "map" of knowns. Some spots, organic compounds, had color, but others did not. In that case, Bassham would spray the spot with a compound thought to react with it to produce an identifiable color. Needless to say, this was a lot of work.

But we're not done yet. Calvin and his coworkers added another element to the analysis: autoradiography. Each spot on the paper chromatogram could be analyzed for its radioactivity using a Geiger-Müller tube with a large window. Sometimes the paper was sandwiched with photographic film, harking back to the days of Becquerel and Rutherford. After waiting for some time, usually 24 hours, they could visualize which of the compounds they separated had isotope and, to some extent, how much radioactivity was there. The chromatograms were sandwiched with film used for X-rays, and as long as there was sufficient radioactivity, and as long as the chromatograms could be protected from cosmic rays, the spots indicated which compounds were labeled, and roughly by how much.

Once the spots were identified and autoradiographed, the team could even locate where in the compound the carbon-14 atom ended up. With some further carbohydrate chemistry, they broke down the compound into individual units and then remeasured the source of the radioactivity. Again, this was tedious, time-consuming, exhaustive work.

The lab was poised to make history. It had a candidate, or "model," photosynthesizer, the green microalga *Chlorella*. It had a long-lived radioactive isotope of carbon, in the form of carbon dioxide, and it could analyze in detail organic compounds in the metabolism of the cells. And the lab had the lollypop. The lollypop vessel (figure 6.2) was designed to be illuminated from either side by powerful incandescent lamps. The openings at top and bottom gave experimenters the means to quickly add radioactive carbon dioxide and, crucially, a means of quickly evacuating the vessel, at the bottom, for assaying where the labeled carbon dioxide ended up.

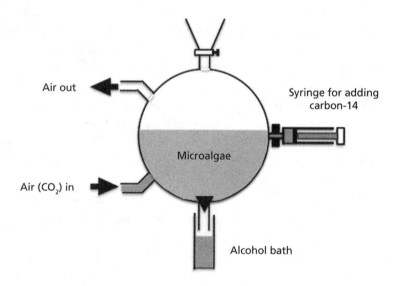

FIGURE 6.2

The lollypop experiment setup. The flask containing the algae, *Chlorella*, was illuminated on both sides, and there were valves to let air bubbles into the culture and to let them out. There was a syringe injection port for introducing the carbon-14, as labeled bicarbonate, and also a valve to evacuate the culture very quickly into a hot alcohol bath, to kill the cells and stop the reactions.

The program by the Calvin team involved a whole series of individual experiments. But the basic experiment had the *Chlorella* growing at a more-or-less steady pace in the lollypop. Air containing carbon-12-dioxide was bubbled in to get them going with photosynthesis. Then the air was turned off, or replaced with nitrogen gas, after which small amounts of carbon-14-labeled bicarbonate (in solution) were added, not enough to perturb the culture conditions but enough that could be assayed at the completion of the experiment. The radio-labeled bicarbonate, a solid powder, was easier to handle than a radioactive gas, and once dissolved, the bicarbonate quickly equilibrated with the carbon dioxide already in the culture vessel; so, in effect, carbon-14-labeled bicarbonate became carbon-14-labeled carbon dioxide.

The culture was exposed to light for varying amounts of time, along with changes to temperature and other environmental conditions. At the end of each treatment, the entire content of the lollypop was immediately flushed into a bath of very hot ethanol. The ethanol destroyed the algal cell walls, leaving an organic-matter soup. The soup was evaporated to a slurry that could then be analyzed for the products of photosynthesis and, hopefully, given the very short duration of the experiment, the very first carbohydrate molecule formed.

This procedure had to be reproduced innumerable times just to get that first product identified. Then, in succeeding trials, the team experimented by sampling the culture at various times with the brief flash, and after turning off the light. The experiments, either with the light on or in the dark, lasted only a few seconds, making the lollypop, with its rapid-empty feature, important. Each experiment required careful procedure, knowledge of carbohydrate chemistry, and technical skills at paper chromatography and autoradiography. And each of those experiments lasting only a few seconds or minutes translated into days of analysis: the two-dimensional paper chromatography, the autoradiography, the identification of unknowns, the chemistry, and the mental work of trying to piece it all together.

What happened? Well, three things. The most obvious spot on the chromatograms with the very short exposures to light, five seconds or less, when analyzed, turned out to be phosphoglyceric acid, abbreviated as PGA. PGA is composed of three carbon atoms with a phosphate group attached (figure 6.3). Given that PGA was the simplest compound

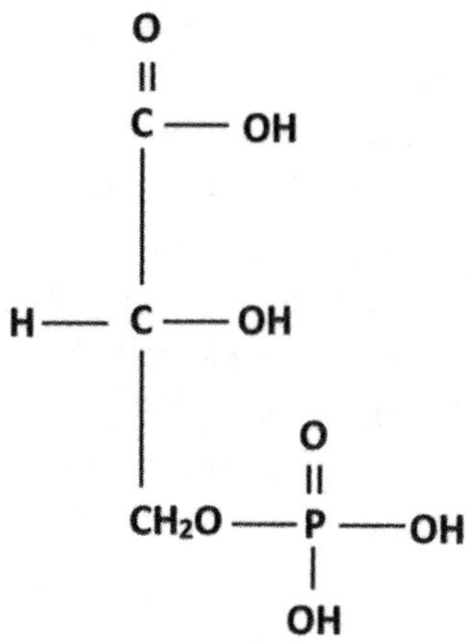

$$
\begin{array}{c}
O \\
\parallel \\
C \text{——} OH \\
| \\
| \\
H \text{——} C \text{——} OH \\
| \\
\quad\quad O \\
\quad\quad \parallel \\
CH_2O \text{——} P \text{——} OH \\
\quad\quad | \\
\quad\quad OH
\end{array}
$$

FIGURE 6.3

The chemical structure of 3-phosphoglyceric acid, with the three carbon atoms and the attached phosphate group—in these experiments, the probable first product of carbon dioxide fixation in photosynthesis.

produced within the shortest period of light exposure, Calvin and Benson suspected that PGA was the elusive first product in the carbon pathway of photosynthesis.

Even though simple, PGA did not identify itself so simply. Overall, sugars with phosphate molecules attached were acidic and didn't move very far on the chromatograms. Calvin's team had to treat these phospho-sugars with an enzyme called a phosphatase that lopped off the phosphate molecules. Then, the spots had to be rechromatographed for better separation of the compounds.

In addition to PGA, there was one very dark spot near the origin of the chromatograms. Treated with phosphatase and rechromatographed, it then moved further than most compounds treated this way. And it moved further than most of the other sugars that were later identified as having six carbon atoms. The dark spot was identified as ribulose-1, 5-diphosphate, or more simply, ribulose bisphosphate, or in shorthand, RuBP (Benson 1951). Benson had to utilize several chemical tricks and transformations to isolate and crystallize these experimental products, but the end result was confidence in their findings.

If PGA was the first product of photosynthesis, how could you get a three-carbon atom from just carbon dioxide? Initially Calvin's team thought there might be a two-carbon molecule that combined with a carbon dioxide molecule to make the three-carbon entity. No such molecule was found.

The second of the three major results provided the clue. In an experiment to see what would be produced after the light was turned off for a few seconds, ribulose bisphosphate, RuBP, plummeted to zero, followed a little later by a falloff in PGA. The third major result was that in the initial seconds of the dark period, before it started falling, PGA increased twice as fast as the uptake of carbon dioxide when the algae were photosynthesizing. The key word here is "twice."

How do we interpret the three findings? Remember, these experimental results were the result of not just one, but many, many experiments, with all the tedium of maintaining stocks of the algae, the two-dimensional paper chromatography, the tricks of chemistry, the autoradiography, not to mention keeping enough carbon-14 available.

RuBP turned out to be the key to it all. Ribulose is a sugar, a carbohydrate with five carbon atoms in a chain. The "bisphosphate" moniker refers to the phosphate groups attached at either end of the five-carbon chain—hence, ribulose bisphosphate. (In those days it was called ribulose diphosphate.)

Bassham credits Calvin with the actual mechanism for reacting a five-carbon sugar with carbon dioxide to make two three-carbon sugars. Calvin himself said it came to him while on his drive home, parked and waiting for his wife to run an errand (Calvin 1989).

Unlike the other sugars identified as intermediates in the pathway by Calvin's group, RuBP is a chain of five carbon atoms, as shown in figure 6.4. The phosphate groups (a phosphorus atom with three oxygen atoms attached) are positioned at carbon atom numbers 1 and 5. Carbon atom number 2 in the chain is different, and has a double bond to an oxygen atom. Calvin realized that a split at number 2 would allow the two phosphate groups each to become part of the split molecule. So carbon-2 should be the site of the split.

The result is that the ribulose bisphosphate splits, forming a three-carbon and a two-carbon sugar, with the two-carbon sugar quickly becoming a three-carbon sugar by the addition, called "carboxylation," of the

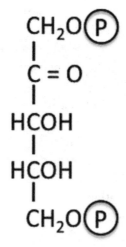

FIGURE 6.4

Ribulose bisphosphate, a five-carbon sugar with phosphate groups attached at each end. The phosphate groups (see figure 6.3), here represented by a circled "P" for simplicity, are attached at carbon atoms 1 and 5. Carbon atom number 2 has a double bond with an oxygen atom.

one-carbon molecule, carbon dioxide. The formation of PGA explains why RuBP declines to zero, and the initial rapid increase in PGA in the few seconds after light is turned off. To a chemist, RuBP (five-carbon) plus carbon dioxide (one-carbon) doubles the amount of PGA (three-carbon). One molecule of carbon dioxide fixed means two molecules of PGA formed from the split of the RuBP. If PGA was the initial product of photosynthesis, it meant that ribulose bisphosphate, RuBP, was the initial acceptor molecule for carbon dioxide in photosynthesis.

There was one further experiment to verify these results: turning off the supply of carbon dioxide. If ribulose bisphosphate was the initial acceptor, its concentration should increase if there was no more carbon dioxide around. The team found that RuBP did increase when carbon dioxide was cut off. In this same series of experiments involving removing the carbon dioxide, PGA would be expected to decrease. And PGA behaved as expected. With no carbon dioxide present, ribulose bisphosphate would not be split and would increase. PGA would not be formed, and would decline.

Thus, the assimilation, or fixation, of carbon dioxide in photosynthesis split the five-carbon ribulose bisphosphate molecule into two three-carbon phosphoglyceric acid molecules (PGA), with PGA being the first product of photosynthesis and the ribulose bisphosphate being the initial acceptor of carbon dioxide. This simplified scheme explained the initial increase of PGA, the decline in ribulose bisphosphate after the light was turned off, and the doubling of PGA relative to the molecules of carbon dioxide taken up in the initial seconds that the light was on. It also explained the results from the experiment in which the carbon dioxide supply was shut off.

How is the carbon dioxide actually "fixed"? Like all biological reactions requiring energy, carbon fixation is catalyzed by an enzyme. That enzyme, now called ribulose bisphosphate carboxylase/oxygenase, accomplishes this feat. Since mouthing that name can tie the tongue, it was shortened to "Rubisco," with humorous asides accumulating over the years, from being put on the covers of cereal boxes to "The Rubisco Kid." In any event, the name stuck.

Calvin and his group were excellent chemists, but not biologists, and they were not too concerned about how all the transformations in the photosynthetic carbon pathway happened or how they were regulated. Rubisco was first hinted at by Sam Wildman at Caltech, who was separating out plant proteins. He called one of them "fraction one protein," because it was so abundant in plant leaves, and also a large molecule. Benson suspected that fraction one protein might be the one catalyzing the fixation of carbon, but he did not remain long enough at the Radiation Laboratory to see his hypothesis through.

Rubisco is a very old enzyme, going back to the initial evolution of photosynthesis itself. Being so ancient perhaps explains some of its quirky and conflicting features. For the enzyme that allows the entry of carbon dioxide into the biosphere, and is responsible for everything that photosynthesis has done for planet Earth, it is quirky indeed.

First, compared to most other enzymes, Rubisco catalyzes the carbon-fixation reaction slowly. Most enzymes turn over biomolecules at a rate of hundreds of thousands a second. Rubisco only converts a paltry ten per second, on a good day. Maybe speed was not necessary in those early Earth conditions. It compensates for not being very quick, as enzymes go, by being plentiful—so plentiful that Rubisco can be considered to be the most common enzyme in the biosphere, the most abundant protein, and something like 10 percent of the mass of all the Earth's biota.

Second, the "o" at the end stands for oxygenase. Rubisco can operate on oxygen as well as carbon dioxide (the carboxylase function), and does so depending on the concentrations of each. If it catalyzes the entry of oxygen into RuBP, it produces not two PGA molecules, but one PGA and another one called phosphoglycollic acid. When phosphoglycollic acid is then metabolized, it produces carbon dioxide! Because this looks like respiration, plant physiologists call this process "photorespiration."

Remember, oxygen is a product of photosynthesis and, in some respects to plants, a poison to be gotten rid of. Scientists are not clear how the oxygenase function came to be, but evolution optimizes only so far as something works—has adaptive value. So we might explain the dual nature of Rubisco as an enzyme that does its job adequately, with a few workarounds. Given when it evolved, along with photosynthetic life maybe 2.5 billion years ago, there was no oxygen in the atmosphere to worry about. As just noted, early photosynthesizers needed to get rid of the oxygen produced; if it got too abundant, they could take it up, just like carbon dioxide, and "idle."

According to Andrew Benson (see Buchanan and Wong 2013), Calvin figured that the fixation of carbon dioxide into organic molecules involved a cycle. This seems to have been based more on intuition than on experimental results at the time, but it turns out he was right. They did realize, however, that no simpler compounds—that is, two- or one-carbon molecules—were found, so it was reasonable to expect that carbon fixation involved regenerating already existing larger molecules, such as the five-carbon ribulose and another in the pathway, sedoheptulose, a seven-carbon sugar. The cycle that ultimately fixes carbon from carbon dioxide into organic matter—that converts three-carbon sugars to five-carbon sugars, in a cycle, the photosynthetic carbon pathway—now bears his name, the Calvin cycle. However, in recognition of the strong contributions from his associates, Andrew Benson and James Bassham (Benson 2002; Bassham 2003), it is often referred to as the Calvin-Benson (CB) cycle, or the Calvin-Benson-Bassham (CBB) cycle.

The CBB cycle is complex, and instead of biochemical detail, we can get the basic premise with some simple arithmetic. Remember that ribulose bisphosphate is a molecule with five carbons arrayed as a chain. Most metabolic intermediates throughout the cellular machinery are six-carbon molecules of various atomic arrangements (rings, chains, etc.), such as glucose, cellulose, and the like. When a carbon dioxide molecule is to be fixed, the five-carbon ribulose splits into a three-carbon molecule and a two-carbon molecule, the carbon dioxide attaches to the two-carbon molecule, and we end up with two three-carbon chains. This is the first product of photosynthesis identified by Calvin's group, 3-phosphoglyceric acid, or PGA.

We now have two three-carbon molecules, and somehow we have to get at least one of those into metabolic pathways, such as for sugars or proteins or lipids, in order to have net carbon fixation for growth. For example, if glucose, a six-carbon molecule is created, its chemical energy can be harvested in cellular respiration, regenerating carbon dioxide and consuming oxygen.

The solution to net fixation of carbon is to go around the cycle a couple more times, or go through three parallel cycles. If we reach this point three times, we end up with a total of three cycles times the two PGAs, becoming six three-carbon molecules of PGA. One of these is sent elsewhere, to those other pathways, leaving five three-carbon molecules. At this point, you can probably see where the cycle is going, since with some nifty chemical rearrangements, catalyzed by enzymes, five three-carbon molecules can become three five-carbon molecules. Doing so, the cycle regenerates the five-carbon molecule, RuBP, used by Rubisco, and sends one three-carbon molecule off to other metabolic cycles in the algal (or plant) cell.

Figure 6.5 shows a simplified version of the overall cycle, for those interested in the pathway details.

All these transformations take energy, and in all cells it is the energy contained in the phosphate bonds that moves things along. The phosphate bonds themselves have to be "contained," so to speak, and that is the job of particular energy-containing molecules in cellular metabolism: adenosine triphosphate, or ATP for short, or nicotinamide adenine dinucleotide phosphate, abbreviated as NADPH. All the transformations in the Calvin-Benson cycle require the energy supplied through these molecules. ATP breaks down to ADP + phosphate, giving up its chemical energy when

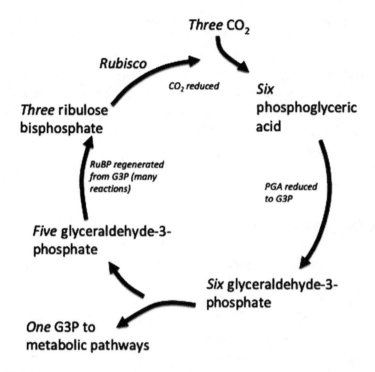

Three CO$_2$

Rubisco

CO$_2$ *reduced*

Three ribulose
bisphosphate

Six
phosphoglyceric
acid

*RuBP regenerated
from G3P (many
reactions)*

*PGA reduced
to G3P*

Five glyceraldehyde-3-
phosphate

Six glyceraldehyde-3-
phosphate

One G3P to
metabolic pathways

FIGURE 6.5

The Calvin-Benson-Bassham cycle, simplified. Three CO$_2$ molecules are combined with three RuBP molecules via the enzyme Rubisco to create six three-carbon molecules of phosphoglyceric acid (PGA). The six PGA molecules are chemically reduced to six glyceraldehyde-3-phosphate (G3P) molecules, one of which is exported for metabolism. The five remaining three-carbon molecules then go through enzyme-mediated transformations to become three five-carbon RuBP molecules, replenishing the cycle. That is, five three-carbon molecules become three five-carbon molecules.

carbon dioxide is fixed into PGA. NADPH breaks down to NADP+ and is used in another step in the cycle.

How are these energy-containing molecules themselves regenerated? That is the job of the light-dependent reactions in photosynthesis. The energy in photons of light excites the pigments (e.g., chlorophyll), splitting water molecules, and the energy evolved from that is used to create, principally, NADPH. When the light is turned on, or the sun rises to a new day, the light reactions kick in, providing the energy ultimately to fix carbon dioxide into organic matter. When the light is turned off, the cycle runs down after a while, because the plant or algal cell runs out of chemical energy. A quasi-independence exists between the so-called "light" reactions, in which solar energy becomes chemical energy, and the "dark," or light-independent reactions, in which that chemical energy is used to fix carbon dioxide.

■ ■ ■

Lawrence started Calvin on the biochemistry of photosynthesis in 1946. Seven years and some 30 scholarly publications later, in 1953, the Calvin-Benson-Bassham cycle became established, unmodified across most plants and photosynthetic microbes. None of the workings of the Calvin cycle could have been achieved without having a long-lived radioactive isotope of carbon, carbon-14.

When the carbon atom of carbon dioxide is labeled, that particular carbon atom can be traced by separating the various intermediate molecules, at particular time points and under various conditions (e.g., light vs. dark), through paper chromatography, and assaying their radioactivities with autoradiography. Identifying the various sugars takes substantial knowledge of carbohydrate chemistry. Then come the deductions to explain what becomes what: what particular intermediate molecule exists where in the cycle, and where does it go? And do the various experimental treatments support the reactions hypothesized? That the Berkeley team was able to deduce the photosynthetic carbon cycle is even more remarkable given the methods available at the time.

Earlier, in the 1930s, Hans Krebs, a German-born biochemist working in England, worked out the first biochemical cycle, predating discovery of the Calvin-Benson-Bassham cycle. The Krebs cycle does the opposite of the CBB cycle, using oxygen in respiration and producing carbon dioxide. The Krebs, or trichloroacetic acid (TCA), cycle occurs in virtually all organisms, both aerobic (in the presence of oxygen) and anaerobic. Krebs demonstrated the cycle that bears his name not using carbon-14, of course, but with manometry, developed by Otto Warburg. Manometry uses slight pressure changes to determine how much oxygen is consumed in a given reaction. Testing various substrates, and how much oxygen was consumed, Krebs worked out the cycle, despite Warburg himself saying that it was a fruitless path of research. Manometry has been used for decades to study respiratory reactions in various animals: sea urchin eggs, pigeon breasts, etc.

We can consider the Warburg apparatus as an analog to using a radioactive tracer. The difference is that Krebs could easily test various known intermediates, whereas Calvin and his group had to discover them. Calvin's group located the position of each carbon-14 atom in the multitude of intermediates, and doing this settled the details of the cycle. They found out, for instance, that carbon-14 occupied the first position in the three-carbon PGA molecule, and that most of the carbon-14 labeled the central carbon atom (C-3) of ribulose bisphosphate. These and other labeling attributes reinforced the team's identification of the pathway.

According to associates, throughout the years when he and his Bio-Organic Group worked out the cycle, Calvin was under intense pressure, contributed largely by a scientific controversy that the Berkeley team faced with researchers from the University of Chicago, led by Edward Fager and Hans Gaffron. (I'm sure competition to be first also played a role.) The culmination of the controversy took place early in their efforts, in 1947, at a meeting of the American Association for the Advancement of Science held in Chicago, where both sides made their case. In the proceedings chapter coming out of that meeting, Fager said, "[our] paper should be considered as a progress report. It is presented in such detail at this time only because of the complete and astonishing disagreement between the results of the work done at this laboratory and that done at the University of California." What was the "complete and astonishing disagreement"? As indicated above, Calvin and Benson identified phosphoglyceric acid as the initial product of carbon dioxide fixation. Fager and Gaffron disputed this finding. They found little carbon-14 activity in these compounds in their experiments.

Fager's lab's methods were entirely chemical, involving cycles of acidification and precipitation, oxidations, steam distillation—methods that were not as precise as the chromatographic methods used by Benson and Calvin. Benson and Calvin, for example, said that any radioactivity might be severely diluted by the large amounts of other plant constituents, and further that Fager and Gaffron's experiments were not of short enough duration to identify the initial products. In the end, Fager and his associates could only conclude a negative—what the initial product of carbon fixation was not. According to them, the initial product was not a carbohydrate, not an amino acid or a keto acid, nor an aldehyde, phenol, or alcohol.

Fager eventually conceded the argument, and Fager and Gaffron, by personal communication, later indicated that they had found appreciable phosphoglyceric acid as an initial product. It was years later, when Calvin, Benson, and Bassham's work held sway and was validated by further work. Fager subsequently withdrew from biochemistry in favor of ecology. After some reeducation, he became a faculty member in marine ecology and biological oceanography at the Scripps Institution of Oceanography in La Jolla, California. Fager was well respected by colleagues and students, and became chair of the department of oceanography at UCSD until his retirement in 1973. On a field trip to Yucatán, he contracted meningitis, from which he never recovered, dying in 1978.

Calvin was 35 years old when he was enlisted by Lawrence to continue Kamen and Ruben's work on the photosynthetic carbon cycle. Three years later he had his first heart attack. Some hinted that Calvin's poor health also hastened Benson's getting fired from the group. Thanks to Calvin's wife, who put him on a strict diet and off cigars, Calvin regained his health and continued a long career. Calvin received the Nobel Prize in Chemistry, solo, in 1961,[3] and thereafter continued research into the biochemistry of photosynthesis. Strangely, he never acknowledged the critical roles played by Andrew Benson and James Bassham, even though they were his coauthors, and often the lead author in the group's publications. Calvin's 1992 autobiography, *Following the Trail of Light: A Scientific Odyssey*, never mentions Benson. Later in his career, Calvin took his research into a prescient direction, looking at the production of oil by plants as a renewable energy source. He died in 1997.

Andrew Benson, after being fired from Calvin's group, spent a short while at Penn State and then returned to California—like Fager, to Scripps. There he continued work on biological systems and, given Scripps's dominant interests, marine biology, investigating the predominance of wax esters as storage products in plankton and pelagic fish. He called these wax esters—complex fat molecules—nature's "starvation insurance," but decidedly not fit for humans. He also had side interests in anthropology, salmon aging, and methanol as a growth stimulator for plants. He died in 2015. James Bassham continued his work on photosynthesis through the 1970s, at one point becoming associate director of the Calvin Laboratory at Berkeley. He died in 2012.

■ ■ ■

This chapter highlights the most important early application of carbon-14: discovering the path of carbon in photosynthesis. Scores of other applications exist throughout the biochemistry of living things. Labeling specific carbon atoms in biochemical cycles allows the biochemist to understand the pathways. For example, if you feed an organism a substance with carbon-14 atoms labeled at a specific point in the molecule, then what happens to that carbon-14 atom? Or, alternatively, how can we find out where each carbon atom came from in a product of metabolism? Called "precursor/end-product" experiments, they are similar in kind to the experiments that Calvin's team conducted. If we know how a specific molecule is metabolized, we can also identify the enzyme involved. At a cruder scale, if we eat chicken eggs whose cholesterol is labeled with carbon-14, does the cholesterol in our body become labeled? We can now exhibit all the cellular metabolic pathways and cycles on one large poster, for reference and for education. Overwhelmingly, these have been worked out with carbon-14.

One recent application deserves particular mention: drug discovery in the pharmaceutical industry. Candidate drugs are labeled with carbon-14 at specific sites in the molecule and then followed through "ADME"—absorption, distribution, metabolism, and excretion, all the phases of body metabolism. Drug discovery in this way is expensive, but gives the most complete picture of the action and safety of specified drugs. "Microdosing" is part of the drug-discovery process. Patients are given small, subtherapeutic doses of a candidate drug to check its behavior in the body before expensive full-scale clinical trials. The extreme sensitivity in assaying that carbon-14 affords finds application here. Drugs, being organic compounds for the most part, have carbon skeletons in which carbon-14 atoms can substitute for carbon-12 at specified points with no change in chemical character.

Many organic molecules occur at very low or trace concentrations, whether in the body, in the environment, or in a laboratory test tube. Or the molecule may be difficult to extract or discriminate from similar molecules, or the errors involved in its analysis prohibit an accurate assessment. Here, tracer methods offer a solution, called "isotope dilution." The essence of the method is to add the molecule, labeled with carbon-14, to the system. The molecules added are only partially labeled, but we know the proportion of labeled to unlabeled compound. Then we allow enough time for the molecule to come to equilibrium within the dynamics of the system. The time period for equilibration can be anywhere from minutes to days, depending on how fast things change. Then, when the radioactivity is assessed, it becomes proportional to the amount of the molecule originally there. When the system comes to equilibrium, the proportion in the various pools is the same, and equivalent to the proportion of the added tracer. The dilution of the label in the system becomes a measure of the total amount of the molecule.

With isotope dilution, we have a powerful technique for measuring the quantity of a substance in a system, whether it is living or not. An example might be adding bicarbonate labeled with carbon-14 to a seawater sample, as I discuss further in chapters 11 and 12. We know pretty well the amount of bicarbonate in seawater, so we have a known ratio between the amount of radioactivity added and the nonradioactive component. A surface seawater sample containing phytoplankton, small zooplankton (usually protozoans), and bacteria might come to equilibrium in a day or two, reflecting the turnover time for this microscopic community. If we take samples to assess the radioactivity at intervals spanning a couple of days, we will find an increasing amount of carbon-14 being taken up and then passed around among the three major components: phytoplankton taking up carbon-14 by photosynthesis, zooplankton becoming labeled by eating the phytoplankton, and label incorporated into the bacteria because they are taking up carbon-14 in waste products produced by both the phytoplankton and the zooplankton. As time goes on, the amount of label recovered in the particulate matter levels off; the system comes to equilibrium, with the carbon-14 taken up in photosynthesis balanced by the loss of carbon-14 through respiration occurring in the three groups. From the amount of label in the particulate matter, compared to the proportion of isotope originally added, we can easily calculate the carbon biomass in this planktonic system. A similar experiment can be done with any number of isotopes. I have simplified the interactions here, and also the ease with which such an experiment can be conducted, but this example illustrates how carbon-14 can be used to understand microbial dynamics.

7

SCINTILLATIONS AND ACCELERATIONS

From across the lab, we must have looked strange, if not crazy. Four legs coming out of the top of what looked to be a top-loading stainless-steel refrigerator, like the old vending machines for bottled soda pop. Two of the legs were mine, and the second pair belonged to a company technician called in to fix our LSC, the liquid scintillation counter. In those days, scintillation counters were refrigerated to reduce electronic noise and background radiation. At the time, I was a beginning graduate student starting to work on a problem in ocean productivity suggested by my supervisor, Gordon Riley. Without a working counter, I might as well have chosen a different research topic.

The company tech found them first, and got himself back upright. "There are the two little buggers." I looked in his palm, cradling two small round pellets. He gently tipped his hand, rolling them back and forth in his palm. These were the radioactive standards for the counter, americium-241 and cesium-137.[1] I gulped. Being a relative newbie, I had no idea that the counter had these radioactive pellets as part of the system. I figured I'd have a fate like Madame Curie. But the tech didn't seem bothered by recovering these bits of radioactivity, necessary to the operation of the counter. And as I later discovered, they weren't very radioactive; they didn't need to be. They were just a stable source of radioactivity to compare with the unknown radioactivity in a sample.

The idea of scintillation counting of radioactive decay events came early in the atomic age, not long after the discovery of radioactivity itself. Like Becquerel's discovery some years before—and frankly, much of science—scintillation counting started with a mistake. Back in the early 1900s,

William Crookes spilled some radium bromide that he was looking at through a zinc sulfide screen. The screen allowed him to observe the fluorescence of the substance directly, using that amazing detector, the human eye. The radium bromide was both extremely expensive and rare, and in his eagerness to recover it all, he picked up a magnifying glass. Using the magnifier along with the zinc sulfide screen, he was surprised to find that the fluorescence became individual flashes of light—scintilla—and each scintilla was a radioactive decay event. Thus was invented what he called the spinthariscope; "spinth" is Greek for spark. The spinthariscope did not last long as a scientific analyzer, since all that was needed was to replace the human eye (and brain) with a photon detector and something that could count automatically. Spinthariscopes, however, can still be found today, marketed as a scientific educational toy. Online they are sold with the catchphrase "See Genuine Atoms Split. $35."[2]

Scintillation detectors can be solid, liquid, or gas, but the most common type uses a liquid. Gas detectors require combusting the sample to a gas, and gas is not easily contained or prepared for counting. Solid detectors are made as a crystal. Solid systems need high-energy radioactivity to penetrate the crystal and produce light. Solid scintillation detectors are only useful for gamma radiation; alpha and beta emitters would never reach a protected solid scintillator. Liquid scintillation is left as the most suitable choice, not requiring combusting the sample into a gas or requiring the radiation to penetrate a solid. Also, most biological samples, for which carbon-14 assumes primary importance, are liquid, or can be used in a liquid environment.

Liquid scintillation counting, introduced in the 1950s, is the most common form of radiation assessment, most commonly for beta-emitting radioisotopes, and therefore especially where carbon-14 is added as a tracer of a biological process. The liquid in scintillation counting is mostly organic solvent, such as toluene, or other "aromatics." Aromatic solvents have chemical bonds in their molecules that enable them to transfer energy. They can transfer radiation from whatever source, and here the source is a beta particle from radioactive decay. The working part of the liquid is the "fluor," or a "scintillator," composed of molecules that are excited by radiative energy. An example is a sodium iodide crystal containing an impurity such as the rare earth element thallium. On receiving energy from the aromatic, the molecules of the fluor are momentarily excited, and when they return to their stable state, they emit light. The light emitted by the fluor molecule depends on the impurity atom itself and the surrounding crystal structure. When working perfectly, one beta emission will result in one scintilla of light. As I discuss below, fluors rarely work perfectly. As a further modification, a chemical detergent is often added to accommodate samples with water in them, which otherwise would not dissolve in the organic solvent.

These three components—the solvent, the fluor, and the detergent—make up the "cocktail," and not one you would want to see on the menu at a bar. When a decay event occurs, the energy is passed around the molecules of the aromatic solvent until a scintillator molecule is hit, at which point a fluorescent flash, a scintilla of light, occurs. A special kind of vacuum tube, a photomultiplier, or PMT, acts like the human eye in the spinthariscope. A PMT senses the scintillas of light and turns those bits of energy into an electrical current. A photomultiplier tube is akin to many vacuum tubes that amplify electric currents. In this case, the detected photon produces an electric current, which is amplified and counted. Modern LSCs use two PMTs positioned opposite each other and coupled electronically. To reduce extraneous noise, only counts from both PMTs are accepted.

The working unit of liquid scintillation counting is the scintillation vial—a small container, 7 or 20 milliliters in volume, usually made of glass or translucent plastic. The radioactive sample—particles, tissue, or another liquid, typically on some kind of support such as a paper disk—is put into the vial, and the cocktail added. Once the sample is loaded into the counter tray, the LSC does the rest (figure 7.1). In fact, most of the workings of the LSC involve moving vials around robotically and getting them into and out of a counting chamber sequentially. A sample in position to be counted descends via a small elevator to a lightless chamber, sealed on top by a shutter and shielded from environmental radiation, again by lead bricks. After a period to allow the cocktail and sample to adapt to the dark, a window opens onto the photomultiplier tubes, which begin detecting the scintillations of light. After a defined interval, programmed into the machine depending on the expected radioactivity, the reverse happens, and after the sample is elevated to its cohorts, the sequence begins anew with the next sample moved into position to be counted. The scintillation

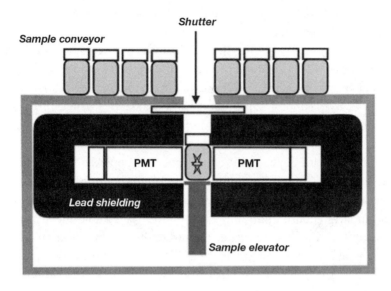

FIGURE 7.1

Schematic of a liquid scintillation counter (LSC) showing the relationship between the conveyed scintillation vials and the sample chamber. At the bottom is an arrangement of photomultiplier tubes (PMTs) with respect to the vial in detecting a scintilla of light (the "star"), illustrating coincidence counting of the sample. Both PMTs have to detect the scintillation within a very short time (coincidental) period (nanoseconds) to eliminate spurious background radiation and record a legitimate event. For automatic external standardization, a pneumatic tube can deliver a standard—a pellet with known radioactivity—to just beneath the vial, which the PMTs can detect.

counts are sent to a monitor, printer, or other recording device. LSCs have capacities for a hundred or more samples at a time.

Why had the company tech and I gone headfirst into the top-loading scintillation counter? One of the trade-offs with the method is that you can never count all the decays or radioactive disintegrations that occur in samples. Photons emitted by the fluor can be absorbed before they are detected by the PMT. The radioactivity can be, as they say, "quenched." Quenching occurs when decays do not reach the fluor because the sample support—the paper disk, for example—is too dense or opaque; when there are too many decays for the photomultiplier tube to count individually; or because of the material itself. Highly colored samples, for example, will be quenched chemically.

One way to deal with quenching is to compare the counts measured in the sample with a standard of known radioactivity, whose radioactivity is subject to the same conditions as the sample in the vial. What the LSC does is to shoot a small standard, a radioactive pellet, through a pneumatic tube, to a position just underneath the sample vial. Radioactivity of the standard can then be counted along with that of the sample. Then the standard is vacuumed away from the sample, and the sample is counted again. The position of the standard pellet beneath the vial acts as if the standard were counted in the sample itself. The difference between the sample+standard and sample alone gives the standard as if it were the sample. Comparing that value with what the standard is known to be allows a calculation of the efficiency of the count, or how much quenching occurs. The sample count divided by the efficiency gives the true, or at least a truer, assessment of the radioactivity in the sample.

For example, let the sample+standard count be 100,000 counts per minute (cpm) and the sample by itself, 30,000 cpm. We know from the manufacturer that the standard count in the radioactive pellet is 50,000 disintegrations per minute (dpm). Then, $100,000 - 30,000 = 70,000$, and $50,000/70,000 = 0.71$, meaning the sample was counted with about 70 percent efficiency. The actual disintegrations, or decays per minute, is then $30,000/0.71$, or about 42,000 dpm.[3] The dpm is the number most commonly reported. Efficiencies for carbon-14 preparations are much higher than 70 percent, typically reaching 90–95 percent.

What happened with the counter in our lab is that, perhaps during a move, the lead pieces shielding the sample compartment and detectors dislodged the pneumatic tubes that brought the standards into position beneath the sample. The standard pellets fell out of the tube and into the bottom of the counter, later to be retrieved by our searching eyes and hands. Our LSC had two standards, one used for carbon-14 and the other for tritium (hydrogen-3).

This method for dealing with quenching is called "automatic external standardization," or AES. The "automatic" refers to the automated way the standard is positioned next to the vial and then taken away. AES is one of a few methods for dealing with quenching, and may not be as accurate as others. Usually a few different methods are used at the same time and compared. AES is used here to illustrate how quenching can be corrected.

■ ■ ■

tick . . . tick, tick. . . . . tick. . . . . . hmmm-hmmm. . . . . tick . . . hums: "We built this city . . ." . . . tick, tick . . tick . . tick . . tick. . . . . . tick . . . (coffee time!) . . tick. . . . tick . . . . . . "on ROCK and ROLLLL" . . . . tick.

In human terms, this is what being a radioactive detector is like. Picture yourself as the detector, using your eyeball to count events, such as through a spinthariscope. With one eye you look through the scope to see the individual events, and with the other, look down at a piece of paper and tabulate the counts. Radioactive decay occurs at random, so measuring radioactivity means statistics. If the sample shows a low rate of disintegration, such as occurs with natural abundances of carbon-14 (see chapter 4 on radiometric dating), then you have two bad choices, sort of like the choice of either going to work or watching daytime TV.[4]

For carbon-14, you can either wait a long time, to get enough counts to compute an average rate of disintegrations per unit time within acceptable error, or you can assay a large sample. Especially with radiometric dating, the size or mass of an artifact cannot be controlled. A bone is a bone. What you find is what you get. Further, the assay may mean destroying, or sacrificing, part or all of whatever you've found, and that found something may be priceless, treasured, or one of a kind. For some artifacts, therefore, dating would not be allowed or possible.

The other choice, then, is to wait long enough to get sufficient counts, occurring at random, to be statistically valid. One or two counts per minute as an average will not be enough to say what the "true" level of radioactivity is, because of the errors inherent to enumerating random events. And there will always be background radiation, from cosmic rays or from local bricks and mortar, even in the most lead-encased detector, as I discussed in chapter 4. If the recorded disintegrations fail to exceed the background radiation, the radioactivity of the sample is, statistically and for all practical purposes, zero.

When carbon-14 first came on the scene, researchers rejoiced that there was a long-lived radioactive isotope of carbon. Being "long-lived" meant that carbon-14 could be used in dating samples, for example, from ancient bones, from the bottom of the ocean, from corals, from carbonate deposits in caves—anything that was once living or created by living things. The half-life of carbon-14 is not super long, but long enough to determine ages of things up to, say, 30,000 to 40,000 years. That time frame goes beyond the beginnings of civilization, back into the most recent glacial age. For biochemical research, where carbon-14 is added, or "spiked," into a sample, you generally recover a lot of isotope, and Geiger-Müller counters, or more recently, liquid scintillation counters, can easily measure the decays. If not, the isotope added can be increased to ensure enough decays occur.

Detecting natural abundances of carbon-14, however, and waiting for a carbon-14 decay while surrounded by other radioactivity, presents a huge problem. With a half-life of about 5,700 years, it is easy to conclude that there are many, many more carbon-14 atoms than actually decay. Indeed, for each radioactive decay event, there are more than a trillion other carbon-14 atoms waiting around. Thus, archeologists, paleoclimatologists, and anthropologists—anyone who had interests in the ancient past—realized the stark limitations in working with natural abundances of the isotope. The amount of carbon-14 in a bone, a piece of cloth, or other artifact, or in a liter of ocean water, is so tiny that even with the superior method of liquid scintillation counting, there simply are not enough radioactive disintegrations, unless you are willing to sacrifice a major part of your shell, piece of cloth, or sediment sample, and unless you are willing to wait long enough for the number of disintegrations to become measurable, statistically. Analysis of a single sample could take weeks, far exceeding the attention span of most scientists and the patience of funding agencies.

Enter Richard Muller, who worked in the Berkeley Radiation Lab under Luis Alvarez, his graduate adviser and mentor. For Muller, the idea that solved the problem of detection of environmental or naturally occurring carbon-14 arose directly from a couple of previous dead ends, or more euphemistically, scientific "detours."

■ ■ ■

The U.S. Navy has always been worried about keeping hidden its submarines, and nuclear submarines (which are inherently noisy) in particular.

The Navy realized, among other security compromises, that a nuclear-powered submarine might leave a wake of radioactivity that could be detected—a signal in the ocean resulting in detection of a larger sort, the submarine itself. In the 1970s, the only way to measure the radioactive atoms in the ocean, remotely, was by means of a new technique called laser fluorescence. After "mulling" over the problem, Muller realized that even if laser fluorescence worked, the numbers of atoms entering the wake of a submarine were too few to measure and too short-lived. One leg of the country's nuclear defense triad was safe in that regard.[5]

Nevertheless, the seed for measuring radioactive atoms directly was planted in Muller's brain. Alvarez then told him about failed efforts at Stanford to use a mass spectrometer to measure carbon-14. Muller thought he might apply his laser fluorescence method to that problem, and set about finding out all he could about carbon-14. In the meantime, however, Alvarez set him another puzzle: quarks. Quarks are the elusive entities thought to make up protons and neutrons. They had never been detected, but Alvarez believed that they might exist freely in nature. He wanted Muller to find these infinitesimal masses at almost infinitesimal amounts in the environment, and suggested using the old 88-inch cyclotron at the Radiation Lab. After an exhaustive two-year-long search, and extending the capability of the cyclotron in new directions, Muller and his colleagues came up empty: another dead end to add to that of radioactive submarine wakes.

A colleague then told Muller about a report he had just read that proposed using the cyclotron in combination with a mass detector, a mass spectrometer, to look for superheavy elements, those that might exist beyond the current limit of the periodic table. Using a cyclotron–mass spectrometer combination would be another way of finding new elements directly. A bit miffed that he had not recognized this particular new direction himself, Muller began to think about what other uses he could put the cyclotron–mass spectrometer to. In one of those flashes of insight that usually occur during routine drudgery—in this case, like Calvin's, driving on his commute home—he arrived at carbon-14. Why not measure carbon-14 directly rather than have the carbon-14, in effect, measure itself through radioactive decay? At home, he began calculating. He checked his numbers over and over again, into the night, realizing that if he was to show them to Alvarez, he had better have them right. The next morning, after hearing Muller out, all Alvarez said was, "Congratulations."

The search for a direct measurement of atoms, the radioactive by-products of fission, plus the search for quarks and developing new and improved uses of the old Berkeley cyclotron, led Muller to the realization that he could use the cyclotron with a mass spectrometer to measure the mass of carbon-14 relative to carbon-12. Cyclotrons had been used since the late 1930s to determine masses of isotopes. But it was left to Muller to come up with the idea that a cyclotron could also be used as a high-energy mass spectrometer—that is, a device able to detect the difference in mass between carbon-14 and carbon-12. Maybe, as he said, we could avoid the "tyranny of Poisson statistics." For those who don't remember their introductory statistics course, the Poisson distribution allows a probability analysis when any individual event, like a radioactive disintegration, is rare, but the number of possible occurrences of the events is very large. In Poisson statistics, the only way to improve the signal compared to noise is to count for a very long time, or be able to get lots of counts. Achieving either of these is unlikely when determining the natural abundance of carbon-14. Richard Muller showed that the problem can be seen as trace element detection instead of trace isotope decay.

Carbon-14 and the more common carbon-12 differ in weight by about 17 percent. The isotopes also differ in how much they will be affected by a magnetic field, but to be affected, they first have to be ionized, or charged. At the time, the methodology of mass spectrometry was most commonly used for a variety of other, usually heavier, elements, using magnets to discriminate the atomic mass differences in isotopes. Isotopes of the elements to be analyzed are blasted with ionizing radiation, and the molecules of the isotopes become themselves ionized. Because the ionized molecules are charged, they become susceptible to magnetic fields, and the magnetic susceptibility of an isotope is proportional to its mass. When the charged ions are then sent through a curved magnetic field and subjected to that field, they will have different trajectories to a detector depending on their mass. The count of the ions reaching the different locations at the detector can be compared, and their abundances therefore quantified.

That was the idea, anyway, but early attempts with the mass spectrometer–cyclotron combination met with limited success, for two reasons.

For one thing, Earth's atmosphere is 80 percent nitrogen gas. And the most common isotope of nitrogen is nitrogen-14, with the same atomic weight as carbon-14. The same masses with different isotopes means, in effect, a huge contamination problem for any mass analysis of carbon-14. Second, mass spectrometry by itself will never have the sensitivity needed to measure carbon-14 at the quantities that exist in nature. Carbon-14 occurs in only one atom in a trillion—that is, one in 10^{12} carbon-12 atoms. The separation and spectrometry would have to be extremely sensitive to be able to quantify the existence of carbon-14 in various kinds of samples.

For the first problem, contamination with nitrogen-14, Muller (1977) considered different workarounds—for example, using gold foil, which would pass carbon-14 atoms but not nitrogen-14, or trying to eliminate the nitrogen-14 altogether. Neither these nor the other solutions he tried gave the necessary outcomes.

The solution to the nitrogen contamination problem is a story in itself; it was worked out, as things often are in science, in conversations in hallways and social events, among attendees at scientific conferences exchanging views.

A physicist at the University of Pennsylvania, Roy Middleton, famous in the 1970s for producing ions of any number of elements on the periodic table, produced the kernel to the solution. Soon after Muller's publication in *Science*, researchers from Simon Fraser University in Canada, led by Erle Nelson, and a group led by C. L. Bennett, at the Universities of Rochester and Toronto, including researchers from the Ionex Corporation, published back-to-back contributions, also in *Science* (Nelson et al. 1977; Bennett et al. 1977). Nelson's group cited a previous undated and unpublished report by Middleton, which said that "no stable . . . negative nitrogen ions are known." Nelson also remembered hearing, secondhand, during a conference dinner, that Middleton inferred that negative nitrogen ions could not withstand acceleration. At least, he could not detect any, so either they fell apart or, less likely, they did not form.[6] Bennett and his colleagues also credit conversations with Middleton during a previous international conference that remarked on the fragility of negative nitrogen ions.

Mass spectrometry uses ions of the element because ions are susceptible to magnetic fields and are affected according to atomic mass. Mass spectrometry commonly uses positive ions. The trick for carbon-14 was to make negative carbon ions from the sample. Any nitrogen-14 atoms that are around, presumably, are not able to stand up to the energy during acceleration and fall apart, leaving only the carbon-14 atoms to be further analyzed. Thus, the major contaminant in the process is literally blown away.

A particle accelerator solves the second problem: the detection of the mass differences in a trace element. A cyclotron is not up to the task of detection of low amounts of carbon-14. Instead, the cyclotron has been replaced with what is known as a tandem electrostatic accelerator; "tandem" here means two stages of acceleration.

After the sample is bombarded with an ion beam—a cesium gun— to create the negatively charged ions, now free of contamination by nitrogen-14, the negative ions are sent to the tandem accelerator. On entering the accelerator, the negative carbon-14 ions are then (in the second stage) converted back to positive ions by stripping off the ions' electrons. At this stage, any other contaminants such as the ion $^{13}CH^-$ (having an atomic mass of 14) are also removed. A few million electron volts in a field are applied, and by the time the ions leave the accelerator, they are traveling at a few percent of the speed of light. By giving the molecules all this energy, any background interferences are removed. The charged particles are then sent past magnets, and with their different susceptibilities to the magnetic field, the particles end up at different locations, different detectors. The difference in abundances of the two isotopes can then be determined by comparing the amounts at the detectors.

The result of these efforts is accelerator mass spectrometry (AMS), a huge advance in the science of carbon-14. By increasing the efficiency of counting, and measuring carbon-14 itself, AMS pushes back the ages that can be dated. Instead of grams of material to analyze, AMS requires only milligrams. The sensitivity achieved, for carbon-14, is equivalent to 10,000 atoms.

The following example shows how this works, and the power of AMS. In terms of radioactive decay, for 10,000 carbon-14 atoms, 5,000 would decay in ~5,700 years. Fifty carbon-14 atoms, just above background radiation, would decay in about 60 years—meaning a long wait for an analysis if researchers had to rely on liquid scintillation! Instead of days in an LSC in front of a photomultiplier tube, samples can be analyzed in minutes.

AMS takes the measurement directly to the carbon-14 instead of waiting for the carbon-14 to reveal itself. AMS can detect three carbon-14 atoms in 10^{16} carbon-12 atoms, a sensitivity adequate for a carbon-14 age of 70,000, and perhaps 100,000 years.

A couple of years ago, Ellen Druffel invited me to visit the AMS facility at the University of California, Irvine (UCI), where she is the Kavli Professor of Earth Science. Druffel, along with Professor Susan Trumbore and Dr. John Southon, are the founders of the W. M. Keck Carbon Cycle Accelerator Mass Spectrometry Laboratory, occupying nearly the entire basement of their building on the Irvine campus. Both Southon and Druffel have been at the forefront of carbon-cycle research with carbon-14 since the early days of analysis by AMS. Southon is from New Zealand, received his Ph.D. from the University of Auckland, and worked at Lawrence Livermore before moving to UCI. Druffel grew up in California, getting her B.A. at Loyola Marymount, in Los Angeles, and then her Ph.D. at the University of California, San Diego. Trumbore received her Ph.D. in geochemistry at Columbia University and has since distinguished herself in terrestrial carbon cycling.

The La Jolla Radiocarbon Lab of UCSD, where Druffel did her graduate research, pioneered many of the developments in carbon-cycle research using carbon-14, such as Hans Suess's identifying what came to be known as the "Suess effect" and Roger Revelle's discovering the "Revelle factor." (I mentioned the Suess effect in chapter 4; I revisit it and introduce the Revelle factor in chapter 10.)

During my visit to UCI, Southon gave me a tour of the components of the AMS system—the cesium gun to ionize the samples, the tandem accelerator, the magnets, the detectors—and then took me, step by step, thorough the procedure for estimating the amount of carbon-14 in proportion to carbon-12 in samples of coral, shells, seawater, wood, sediment, charcoal, bone, ancient wine, or whatever has carbon in it.

The procedure for preparing the samples is what in chemistry is called an oxidation-reduction sequence, similar to the analysis that Sam Ruben applied to the very first sample containing carbon-14. Oxidation-reduction means oxidizing the organic material to carbon dioxide and then reducing the carbon dioxide to pure carbon. So, in the first step, the samples are burned, or oxidized—evolving carbon dioxide gas, and capturing it.

Then, the captured gas is reduced with a metal catalyst to drive off the oxygen, forming the pure carbon compound graphite. Graphite is what Martin Kamen used as a target in the discovery of carbon-14 (chapter 1). At this stage, nitrogen-14 becomes a contaminant in the graphite, to be removed later.

The few milligrams of graphite sample are glued onto metal disks. The metal disks are the targets, bombarded by a cesium gun, producing negative carbon and nitrogen ions. The ions are then accelerated, and because the negative nitrogen ions are so unstable and easily broken apart, they are lost. After getting rid of the nitrogen-14, the electrons are removed from the carbon ions, making them positively charged. The positively charged carbon ions are accelerated again to ground potential (no charge), and after further focusing by magnets, the ions are directed to detectors that discriminate on the basis of mass, in this case carbon-12 and carbon-14, and also carbon-13. It is seemingly a complicated process, but after preparation of the graphite sample it only takes about 10 minutes from beginning to end. A schematic of the Keck AMS system is shown in figure 7.2.

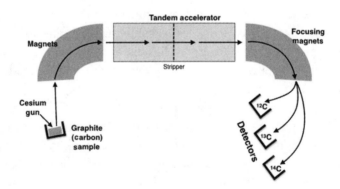

FIGURE 7.2

Simplified schematic of an AMS radiocarbon system, showing the various components for analyzing samples for carbon-14. The UCI Keck AMS is about five meters by six meters, filling a large room. The charged carbon ions produced by the ion source (a cesium gun) are organized by the injection magnet and passed to the tandem accelerator. After the ions are accelerated, they are focused by magnets and eventually detected, by mass difference, as ^{12}C, ^{13}C, and ^{14}C.

Muller had worked out a small version of the AMS that could be purchased and used by individual laboratories and universities. As the method developed, however, it became prudent and more efficient to have a small number of AMS facilities to which investigators would send samples for analysis. The Keck facility at UCI is one of several AMS laboratories around the United States[7] and the world devoted to carbon-14 dating and receiving datable artifacts, seawater, and the like. Most researchers using carbon-14 dating, I suspect, have never looked at an AMS up close. Analyses run about $300 each.

Muller reports that he was invited to give a seminar at the Lamont-Doherty Geological Observatory (as it was known then) in the 1970s, and perhaps I just missed his visit with my arrival there in 1977. Wally Broecker, a pioneer in the science of carbon-14 (see chapters 9 and 13), told Muller many years later that he had made a bet with a colleague for $10 that the cyclotron–mass spectrometry idea would not work—a bet he paid off a few years later. By that time, he was pursuing an AMS facility at Lamont, excited about the prospects for oceanography, paleo-oceanography, and climate studies.

Both liquid scintillation counting and accelerator mass spectrometry were developed after the initial applications (the photosynthetic carbon cycle, carbon-14 dating) of the newly discovered isotope of carbon, carbon-14. Use of scintillation counters began in the 1950s, while AMS did not mature as an assessment method until the mid-1980s. In the next chapters I tell the stories of the follow-on studies: ocean fertility, which relies on liquid scintillation; and ocean circulation and climate history, which utilize AMS.

8

THE SHROUD OF TURIN
AND OTHER RELICS

I remember a comedy bit performed by Father Guido Sarducci (aka Don Novello). Many of his routines are on YouTube, but I can't find this one. So if my memory is in error, I'm to blame. Father Sarducci said he was called to investigate a miracle where every year at a certain time, a spiritual painting bled. Pilgrims came from all over to witness the miracle, pray over it, and otherwise pay their respects. Father Sarducci was asked to investigate, and when he did, he took a close look at the painting, and touched the "blood" with his finger. He looked at it, and in an inspired moment, tasted what was on his finger. The blood was actually ketchup. Not missing a beat, Father Sarducci proclaimed a miracle that a painting could bleed ketchup. I think of Father Sarducci's reaction to the miracle when I look at the story of the Shroud of Turin.

The Shroud of Turin is a linen cloth, roughly four meters long and one meter wide, with the image of a man, both front and back (figure 8.1). It first appeared in 1353, in Lirey, France, and created controversy from the beginning. The first unfortunate happenstance occurred when its discoverer, Geoffroi de Charny, died shortly thereafter, in 1356, at the Battle of Poitiers, one of the engagements in the Hundred Years' War between England and France. The local bishop called it "cunningly painted." Nevertheless, while in Lirey, the shroud did its job of attracting pilgrims and adding to the church's coffers, while at the same time engendering dispute and concern. Amid the quarrels, in 1390, Bishop Pierre d'Arcis called it a forgery, but did not name the forger. Much later, in 1543, John Calvin, no less, denounced it on the basis of biblical inconsistencies. For example, the gospel writers make no mention of a covering sheet or a likeness produced on it.

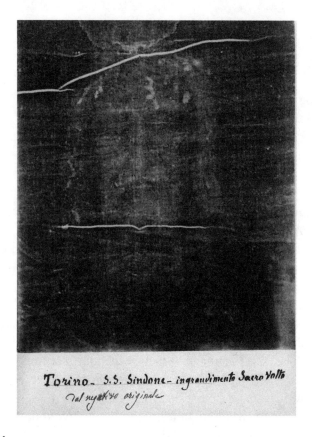

Torino- S.S. Sindone- ingrandimento Sacro Volto
dal negativo originale

FIGURE 8.1

Secondo Pia's photograph of the Shroud of Turin, 1898. (Wikimedia Commons)

Through all these disagreements, and perhaps because of them, the shroud attracted attention over the years. In the mid-fifteenth century, it was sold to the Duke of Savoy, who moved it to Chambery in the south of France. While there, the shroud was partially scorched in a fire, and the burn marks remain visible today. When the Savoys moved to Turin, they took the shroud with them, and there it has remained.

In the late nineteenth century, when Secondo Pia photographed it, the shroud attained new prominence. The photographic technology available at the time required long exposures and very bright lights. Somehow the shroud escaped any damage from Pia's lamps. When Pia developed the film, he was surprised to find a positive image, meaning that the image on the shroud itself was a photographic negative. The photographs revived interest in the original image, and how it came to be produced. To believers, the Shroud of Turin is the burial cloth of Jesus after he was taken down from the cross. According to the story, his body was laid on one end of the cloth, and the other end folded up from his feet to his head, producing both front and rear images. The Shroud of Turin is probably the most intensively studied artifact in human history.

Just about every Catholic church with any pretensions houses a collection of relics, bones, and other handiwork from the saints. I have visited a few churches and seen the reliquaries, if not the actual bones, of St. Mark in Venice, St. Nicholas (from whom our Santa Claus is derived) in Bari, at the heel of Italy's boot, and relics in a family chapel in Umbria. St. Mark's relics were spirited out of Egypt in the ninth century, as portrayed in the mosaics above the portals to Basilico San Marco in Venice. Venetian merchants packed them under some pork products, counting on the revulsion of Muslim customs agents in Alexandria to get them past, and then out of the country.

St. Nicholas rose to fame in Turkey in the fourth century for, among other miracles, giving gifts in secret. Among the stories, St. Nicholas was said to have tossed purses of gold coins through the windows of three daughters of an acquaintance, the coins to serve as dowries, which the father could not afford. St. Nicholas's bones were believed to produce a health-giving vapor. The relics were stolen from his burial place in Lycia (Turkey) in the eleventh century in much the same manner as with St. Mark's, with visiting Barian merchants breaking into his tomb, packing up his bones, and making their getaway, hotly pursued by the townspeople. They subsequently named their church in Bari the Basilico de San Nicola.[1]

One would think that stealing, or at least depriving the rightful owners of their relics, might not be the best example of Christian tradition. For religious relics, money seems to have eclipsed purity of belief in those instances. For both the relics of St. Mark and St. Nicholas, the goal was economic—producing more pilgrims and visitors—and the relics enhanced the prestige of the cities involved, Bari and Venice. The Shroud of Turin was different, in that it was not as tangible as a bone from a saint, but an image, and potentially a powerful one. If it was the burial shroud of Jesus, it had to have assumed much greater reverence, even though in the early days after its discovery, and despite controversy, it enhanced the visibility and coffers of the church in Lirey like any other relic.

As mentioned above, since the mid-1600s, the shroud has been housed in Turin, in a chapel built for it, the Chapel of the Holy Shroud. By the early part of the nineteenth century, the last of the Savoys, Umberto II, was the shroud's owner. During World War II, the shroud was sent to a mountain monastery in the south of Italy for safekeeping. After the war and Umberto II's banishment from Italy, the church assumed ownership. After various incidents (including arson), the shroud has been on display only intermittently, currently requiring an appointment for a viewing. The shroud is kept in a sealed case, with controlled temperature and humidity, under bullet-proof glass.

About 30 years ago, in 1988, three laboratories independently dated the fabric of the shroud using carbon-14 and accelerator mass spectrometry. Interest in carbon dating the shroud arose not long after the invention of accelerator mass spectrometry (AMS). In previous years, dating the shroud was prevented by the large quantities of sample required by counting decay electrons (radioactivity) to get a carbon-14 date. As pointed out in chapter 7, AMS requires only milligrams of material because the method assesses the mass of carbon-14 instead of measuring its decay. Whereas decay counting would need a piece of cloth the size of a handkerchief and maybe months of counting, with AMS, a piece of the shroud only the size of a postage stamp need be sacrificed (Taylor 2016). AMS made it possible to find a date for the cloth, or more specifically, the harvesting of the flax plants that produced the linen woven into the cloth. Since it is the most famously examined religious artifact in the world, dating the shroud could have been a boon for AMS research. However, dating the shroud might also shine a light on the less desirable aspects of the dialogue between religion and science, between data and beliefs.

The first concerns over dating the shroud came not from those with an interest in religious artifacts, but from the scientists. Those who pioneered AMS wondered whether the method was mature enough to take on such a potently controversial task. So, while efforts to arrange a sampling of the shroud linen began in the early 1980s, it was not until

1986 that an agreement was reached. Even then, only three laboratories were to be allowed to analyze the cloth, instead of the seven suggested by the scientists. The smaller number of laboratories meant that an outlier analysis would throw the entire dating procedure into question. The scientists' fears were well founded. An earlier exercise to compare laboratory methods using AMS for carbon dating of cloth had one date markedly different than the others for the sample tested.

After a few years of discussion and preparation, in 1988, three small samples from the cloth were taken to date the flax plants used for the linen. The samples taken were planned well in advance after consultation with the shroud's curators. The actual cutting of the cloth was supervised by the church. A member of the British Museum, witnessed by the Archbishop of Turin, placed the three shroud samples into numbered stainless steel containers. Along with the shroud samples were three other linen "control" samples, for which dates were known independently: from a Nubian tomb, from the wrapping of a mummy from the time of Cleopatra, and from a cloak buried with St. Louis of D'Anjou in France.

The vials were distributed to three of the most respected AMS facilities at the time: the University of Arizona in the United States, Oxford University in England, and the Swiss Federal Institute of Technology (ETH) in Zurich. The test was "double-blind": neither the analysts nor those distributing the samples knew which containers contained the shroud material and which contained the control samples. The question for carbon-14 researchers was whether the flax used to weave the linen fabric could be dated to the first century CE.

The much anticipated results? The three laboratories presented dates based on carbon-14 within one statistical standard deviation of each other.[2] For all intents and purposes, and within measurement error, the dates they arrived at were the same. That was a big "whew!" because an outlier would probably have invalidated, or at least brought into question, the resulting dates. An outlier among seven independent analyses could have been explained, statistically, but not one out of three.

The mean date that the three laboratories arrived at was 1325 CE, with an error (standard deviation) of 33 years (figure 8.2). Statistical probability for that standard deviation allows that there is a 95 percent probability that the date for the flax harvest occurred sometime between 1259 and 1391 CE

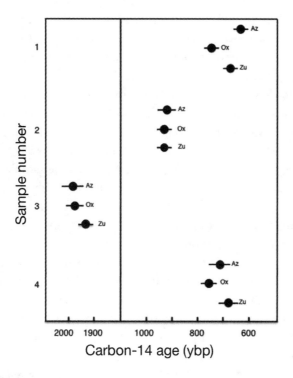

FIGURE 8.2

The results from carbon-14 dating of the Shroud of Turin. The samples are labeled by laboratory: Az for Arizona, Zu for Zurich, and Ox for Oxford. The shroud samples are number 1; numbers 2, 3, and 4 are the controls (after Damon et al. 1989).

(essentially, $2 \times 33 = 66$). The shroud was initially announced in 1353 CE, meaning that the AMS dates were near to the time that it came to be known.

The carbon-14 dating results did not diminish or dissipate the controversy surrounding the shroud—far from it. Almost immediately, there were issues raised, most commonly about contamination of the cloth that would have increased the carbon-14 values and therefore have given erroneous dates—that is, dates that were too recent. The cloth samples were cleaned, and if the cleaning procedures were the same for each laboratory, a case might be made that contamination was not properly removed and the dates, though consistent, were wrong. However, each lab had its own and differing cleaning procedures, so that any contamination would be differentially treated and would only increase the spread of the dates analyzed. One can work out that the shroud would have to be almost overwhelmingly contaminated with modern carbon to bring the date from the first to the fourteenth century.

When the samples were removed from the vials for analysis by the AMS scientists, those from the shroud were recognizable because of the distinctive weave in the linen. Ways of treating the samples to disguise them before distribution were considered, but rejected because of the damage such treatments might cause. Identifiable samples, however, worked in favor of removing the criticisms. For example, there could not have been replacement with other material prior to the analysis. In any case, all the samples were analyzed without consultation among the labs; the analyses were carried out independently. Some of the analysts had actually hoped for a result consistent with the story of the Crucifixion, and the fact that those analysts agreed with the dates produced speaks to the integrity of the entire process. Interestingly, microscopic analyses of a sample of the shroud had been carried out some years previously by the (now deceased) microscopy specialist, Walter J. McCrone. He identified traces of iron oxide in linen fibers taken from the shroud's image. As later reported by Harry Gove (1999),[3] McCrone said, "I have some good news and some bad news. The bad news is that the Shroud is a fake. The good news is that no one will believe me."

The image on the Shroud of Turin is thought by some to represent an energy release at the moment of resurrection. Although such an explanation defies the laws of physics, like the bleeding ketchup in the painting, that's OK, because it is a miraculous occurrence outside physical laws. Ultimately it comes down to faith, which cannot be analyzed with science or carbon-14 dating.

■ ■ ■

Although carbon dates had been accruing for many years, the Shroud of Turin put carbon-14 dating in the public eye. Archeologists had begun using carbon-14 in the late 1950s, as a clock that opened the door to time-mapping recent history, the history of civilization, and recent climate change. "Recent" in geological terms means thousands of years.

Most geological time periods are recorded as millions of years. The development of carbon-14 in age-dating paralleled that of other radiometric clocks, such as uranium decaying to lead, that allowed dates going back to the formation of the Earth.

Whereas before there were the odd written records of dates, now the entire Earth's past was open to chronological mapping. Radiometric dating inspired other ways of dating Earth's past. I have already discussed the annual growth rings in trees (chapter 4), notably the extremely long-lived bristlecone pine trees of California's White Mountains. These amazing trees have lived for 4,000 years, and are used to calibrate carbon-14 ages more accurately. The use of DNA has developed as a way to understand the divergence of various biological groups, showing the history of life, not just the history of humans. The cosmic microwave background—the faint ripples of the Big Bang, the Big Bang's afterglow—gave independent evidence for the age of the universe. The development of radiometric dating using carbon-14 represented one important theme in the entire prospect of creating chronologies, and therefore understanding the processes that influence today's world. Science now operated not only in the present, but on an historical foundation.

Carbon-14 created a "chronometric revolution" that began in the 1960s. "Carbon dating" entered the vernacular, and anything and everything older than, say, a hundred years could be given a date of demise, whether it was the grain used in that "old" bottle of whiskey, the harvest of grapes for a bottle of wine, the shells of marine organisms, or older artifacts from prehistory.

The sciences of oceanography, geology, and astronomy differ from other, so-called "bench" sciences in that researchers cannot conduct an experiment, let alone one with an experimental control. In geology, for example, the experiment, or change, has occurred over previous eons, and we can only see the results. One analogy might be looking at a crime scene from a crime committed without witnesses. To solve the crime, the investigator has to explain what happened using various clues, such as fingerprints or any bits of hair or clothing left behind. Another might be the debate about the sound of a tree falling in a remote forest with no one around to hear it. Scientists could investigate whether, when trees fall, they make a sound, and observe the condition and position of said tree, to conclude that, yes, it did fall, and lab experiments on other trees show that, yes, they generate sound waves. Geologists and, more recently, archeologists look for the fingerprints of past processes to understand how things came to be.

Historical geology is a good example. Geology had its beginnings in England, with well-to-do gentlemen amateurs combing the rock formations and cataloging the layers. Most of the names of the geologic eras come from Devon (the Devonian) or Cambria (the Latin name for Wales; the Cambrian). The Silurian is named for a Celtic tribe. Subdivisions are named for towns or the locations of fossils. The only rule that could be applied was that in an ideal rock formation, shallower layers and fossils had to be more recent in time than deeper layers, the so-called "law of superposition" set forth by Nicholas Steno, a seventeenth-century Danish scientist. The chronology described was all relative, using the qualitative terms "older" and "younger." Perhaps the only rough age that could be obtained, and only for more recent times, was from the accumulation of sediment in lakes, assuming that it was constant over time. For some lakes, sediments accumulated as annual layers called "varves" (as I mentioned in chapter 4 and consider again in chapter 13), and these were used to get a rough idea of dates.

Chronologies are important. Human history requires a reliable sequence of events to understand the nature of development and migration. How important is climate? The landscape? What is the societal response to a historical event? Answers to these questions can only come from a chronology.

A rough chronology similar to that of geology existed in archeology before carbon-14 dating. Some ideas were possible based on the characteristics of human artifacts and monuments, and were hotly debated. For example, a central idea of archeology was that progress in civilization and ideas radiated from the Near East, from the area around the Nile River and from Mesopotamia. Human innovation happened only once and in one area, the Middle and Near East, and over time spread west into Europe and east to Asia, and even as far as the Meso-American civilizations of the Incas, Mayans, and Aztecs. According to this way of thinking, the Egyptian pyramids were the most ancient of human monuments, dating to 3000 BCE. Other monuments with similar cultural function in

western Europe were believed to be more recent and, in effect, dependent on the cultures that had developed in the Middle East. The idea that
culture spread, or "diffused," from a single source proved seductive to
all archeological study in the early twentieth century. Carbon-14 dating
turned the cultural diffusion idea on its head.

Archeologists group innovations into three categories: monuments
to the dead, astronomical observatories, and metallurgy, beginning with
copper (Renfrew 1973; Thorpe 1999). Carbon-14 put these categories
into different chronologies, and in their proper spot in the development
of human civilization. Stonehenge, on the Salisbury Plain in southern
England, exemplifies the reordering of chronologies.

Stonehenge was carbon-14 dated to be about as old (dating to 2500
BCE) as the earliest monuments in the Middle East and Mediterranean,
and other analyses suggested that the site had been used for ceremonial
purposes a half-century before that. As I pointed out in chapter 4, the
Egyptians kept lists of their kings going back to about 3000 BCE, and
Libby therefore had independent dates for his initial carbon-14 analyses.
The aboriginal peoples of England and western Europe had no known
written records. Nevertheless, Stonehenge was now dated to be as old as
the pyramids. Similarly, copper was being smelted in the Balkans before
there are records from the Mycenaeans (the last phase of the Bronze Age)
in the Aegean.

It is not that the dates for Egypt and Mesopotamia were necessarily
wrong. The dates for Egyptian pharaohs were not in question. It is that
the dates for monuments and artifacts in northern and western Europe
were now found to be several centuries earlier than had been assumed by
the "diffusion" of cultural ideas from the Mideast. The cultures of Britain,
France, and Central Europe must have developed independently from
those in Mesopotamia. They could not have had any contact with, for
example, Mycenaean, Nile, or other Aegean civilizations, but at the same
time established similar human attributes.

Carbon-14 dating ushered in a "New Archeology," a revolution. Instead
of a one-time innovation in human culture of honoring the dead, making
metal implements or ornamentation, or figuring out when to plant crops
or understand ocean tides, all of these happened separately, in geographically unconnected human populations. Missionaries did not carry their

religious memes and megaliths to Europe from the Mediterranean; each group developed its own. That this can happen in multiplicity is illustrated by the monuments not only at Stonehenge, but in Malta, and as far away as Easter Island.

If there was not one source for the developments in the Neolithic and Bronze ages, how did they happen in isolation? How did the architects of Stonehenge figure things out on their own without learning from others? Carbon-14 dating cannot answer that question. It can only provide the "when" and the sequence of major developments in memorializing the afterlife, metallurgy, and star gazing. Even though it is just the "when"— the chronology, the sequence of advances in human prehistory—it narrows the possibilities and provides for the intellectual nourishment needed to advance ideas.

The revolution in thinking that followed from carbon-14 dating, as always in science, was not without its detractors and those who refused to see their mental edifices crumble. The concerns expressed by those who refused to accept the results from carbon-14 dating had a useful purpose, however. There was a sliver of truth to their criticisms, and that was part of Hans Suess's second revolution in dating that I covered in chapter 4. Recognizing that there were variations in the carbon-14 years relative to calendar years improved the precision of the dates.

■ ■ ■

Presenting a seminar to geochemists has been a rite of passage for graduate students, postdoctoral fellows, and faculty alike. In their search for understanding, geochemists, at least those who study the ocean, are hard on each other in seminars and conference presentations, always testing the limits of politeness. Perhaps coming from a somewhat staid Canadian university, following more of a British system of graduate education, had something to do with my surprise when I arrived at the Lamont-Doherty Geological Observatory and attended my first geochemistry seminar. Despite the seeming rancor, I did find that for the most part, the participants respected each other; in only rare instances did things get personal. These differences and controversies, however, paled in comparison to the fierce and toxic disputes surrounding the peopling of North America.

Artifacts, particularly arrow points of a distinctive style, were found near Clovis, New Mexico, in 1929. When they and their associated artifacts could be dated using carbon-14, in the 1950s, they were found to be 13,000 years old, the earliest human relics found in North America. The objects comprised what came to be known as "Clovis culture," or the Clovis people, and artifacts began to show up all around the United States. The timing, 13,000 years ago, is just before the Younger Dryas cold snap (see chapter 13). The Clovis people entered a continent with abundant game—mastodon, mammoths, sloths, giant beaver—and put their hunting skills to good use. Clovis became the accepted paradigm for North American archeology, so much so that any conflicting evidence was, at the most generous, disregarded and, at worst, accused of being "bad science." The "bad science" moniker won out. Graduate students from the non-Clovis school of archeology could not get jobs. Sometimes professors found their positions with their universities in peril. Peer reviews of grant proposals contained large doses of malevolence and vindictiveness.

In the 1970s, another site was uncovered 10,000 miles to the south, in Monte Verde, Chile, with earlier dates than the Clovis. Carbon-14 dated the artifacts found there to 1,000 years earlier than those in North America. Then came the critical, almost unthinking, negative response from the "Clovis first" side. They refused to accept these new findings. South American scientists have always bristled at a science dominated by "El Norte," and this was one more example.[4]

Carbon-14 dates have been found for many other archeological sites in recent years, with dates even the Clovis people would have considered ancient. Other than Monte Verde, Chile, there are caves in Oregon with artifacts older than 14,000 years before the present (ybp). Sites in Pennsylvania (16,000 ybp), Texas (13,000–15,000 ybp), the Florida panhandle (14,500 ybp), and the Yukon (24,000 ybp) also predate Clovis culture. Although no human bones have been found at these older sites, evidence of butchering, arrow points, and stone implements have been uncovered, and a few sites have coprolites (fossilized human excrement) that can be dated. The "Clovis first" paradigm has caved. It has pretty much gone from dogma to doghouse.

Yet carbon-14 data, although they have helped, have not dispelled doubts about the peopling of North America. Much of the problem centers around the nature of archeological research itself. You find an artifact that can be dated using carbon-14, but it is one of a kind and cannot be validated with other sites, or most other radiometric isotopes. The artifact's position in the environment, whether it might be contaminated with other material of different ages, and whether it can be properly cleaned leave openings for those who would question any results with which they disagree on other grounds.

Related to the issues surrounding the peopling of North America was a find in 1996 on the banks of the Columbia River in the town of Kennewick, Washington. A skull had been found by some students, who passed it along to the county coroner, who then gave it to James Chatters, a local archeologist. When he and the coroner went back to the site, they uncovered more bones, making nearly a complete skeleton. All Chatters's measurements of the skeleton, which he called Kennewick Man, suggested to him a Caucasian. A facial reconstruction of the skull had Kennewick Man looking like Patrick Stewart (also known as Captain Jean-Luc Picard from *Star Trek: The Next Generation*). At the same time, Chatters had samples from the skeleton dated using carbon-14, and the lab told him the bones were between 9,200 and 9,500 years old. Needless to say, the possibility that Europeans were present in North America that early made for some outrageous headlines. Here was one person, one archeological instance, spun into European invasions, separate waves of arrival of Mongoloids and Caucasians, ancient race wars, and even a revenge story involving Christopher Columbus.[5]

Kennewick Man's provenance, whether he was Native American, European, or even an ancestor of a northern California religious group claiming descent from Germanic tribes, roiled the courts for more than a decade, during which time his bones were prevented from further examination and research. Should the bones be interred in accordance with Native American religious beliefs? The Army Corps of Engineers sided with the Native Americans, sequestered the remains, and even covered up the site, preventing further excavation. The one thing the various combatants agreed on was that Kennewick Man was a huge find.

The case was finally settled in 2013 when methods had been developed that could analyze the traces of DNA found in the bones. Kennewick Man was definitely a Native American, related to tribes such as the Umatilla that resided in the area of Washington where the bones were found.

The resemblance to Patrick Stewart or any other Caucasian seems to have been the result of cultural bias. To Native Americans, Kennewick Man looked more like Chief Black Hawk (painted in 1833) or the chief's direct descendant, the Olympian and football player Jim Thorpe. Kennewick Man was finally buried at an undisclosed location somewhere in Washington, witnessed by Umatilla tribe members.

Carbon-14 dating for Kennewick Man illustrated some of the difficulties and uncertainties with dating archeological artifacts (Taylor et al. 1998). The radiocarbon laboratory at the University of California, Riverside completed the initial analysis for a bone from the little finger—more specifically, the collagen in the bone—and sent it to the Center for AMS at Lawrence Livermore National Laboratory for a carbon-14 value. The uncalibrated age of the bone was 8,410 +/- 60 years. As discussed in chapter 4, calibrating this date from carbon-14 to solar years meant adding another 900 years. Because Kennewick Man was discovered on the banks of the Columbia River, it was assumed that his diet, like that of his modern descendants, consisted significantly of salmon, which, before coming up the Columbia River to spawn, had spent previous years roaming the Gulf of Alaska, or more generally, the North Pacific.

Like all fish (see chapter 10), salmon have a "reservoir effect"—carbon-14 values that show them to be older than current values in the atmosphere, and often in the water they swim in. The Pacific Ocean correction is estimated at 750 years, and it is further estimated that the fish would be 70 percent of Kennewick Man's total dietary intake. Thus, 530 years are subtracted from the 900, giving an age in the range (considering the errors involved) of about 8,800 years before the present.

Dating artifacts always presents uncertainties, whether from the imprecision of the assay, from irregularities in the calibration curve (currently IntCal13), or because of the condition and nature of the artifact itself.

■ ■ ■

Many other carbon-14 dating analyses exist—less famous than the Shroud of Turin, the peopling of North America, or "anomalous" skulls, but no less important. These analyses have opened windows to the past and to other processes that no one might have thought existed. I will mention a few of these to show how broadly carbon-14 dating can be applied. Indeed, dating artifacts or anything else older than, say, a hundred years has now become routine and referred to without even mentioning carbon-14. Carbon-14 dating can settle a whole variety of arguments.

One well-known instance involved one of the more remarkable discoveries of the last century. In 1991, two Germans, hiking in the Italian Alps near the Austrian border, discovered a human corpse partially encased in the ice of a glacier. They cut the corpse out of the ice, along with some artifacts entombed with it, and transported the corpse and artifacts, still frozen, to the Research Institute for Alpine Studies in Innsbruck. The corpse came to be known as Ötzi ("Eastie") for the place it was found. Fortunately, AMS was available to carbon-date Ötzi. Relying on radioactive decay would have involved unacceptable losses to Ötzi's corpse, as well as his tools and weapons.

Because of innate variability in the carbon-14 calibration (chapter 4, and referred to earlier), the AMS dates could only be accurate to within 200 years, between 5,100 and 5,300 years ago. Once the date of Ötzi's demise had been established, most of the analyses concerned the nature of his tools, his stomach contents, an assessment of his health, and most interestingly, his last hours. Researchers have surmised that Ötzi was critically wounded with an arrow, followed by a blow to the head. Ötzi was very probably murdered.[6]

More recent carbon-14 ages and applications rely on the time stamp of the nuclear tests in the 1950s (alluded to earlier, in chapter 4). These include an analysis of ivory, the growth of *Nautilus* chambers, and finding the oldest vertebrate animal.

An international treaty signed in 1989 prohibits trade in ivory. One argument made by ivory merchants is that the ivory they sell comes from ivory stockpiled from before the time it became illegal to market. They convince their customers that their ivory was recycled from older caches of elephant tusk. Carbon-14 analyses by researchers at the Lamont-Doherty Earth Observatory, using ivory from various government seizures, show this to be a deceptive sales tactic. Carbon-14 from nuclear testing in the 1950s and early 1960s entered the atmosphere; it is taken up by plants that the elephants eat and then gets incorporated into the tusks. Those nuclear tests took place before the elephants in question were born, but a small residual amount stayed with the plants in the elephants' diet. There is a part of the tusk that is newly formed, and analyzing the ivory in this newly formed tusk gives the time of death. That death, contrary to what merchants might advertise, is recent, within a few years of sale, indicating that most ivory on the black market today comes from poaching.

Most of us have birth certificates, a document recorded by witnesses to our entry into the world. Most animals in the wild do not, although age distribution and the ages of animals themselves are important for conservation efforts and fisheries management. The age of most fish can be estimated by looking at bones in the fish's ear called otoliths. The otoliths become larger as the fish gets older, and have growth rings, just like tree rings. The age distribution in a population of cod, capelin, or halibut can then be used to manage the fishery. For example, if the age distribution is skewed toward juveniles, it is a sign that the fishery is in trouble.

Sharks, being cartilaginous, do not have otoliths and cannot be aged this way. However, they do have something called an "eye lens nucleus," which has crystallized proteins. The proteins are stable and formed during prenatal development. Thus, a carbon-14/carbon-12 ratio in the eye lens nucleus represents a value at age zero for a shark, and the ratio declines by normal radioactive decay as the shark gets older: the older the individual, the lower the ratio. When Greenland sharks were caught and sampled, the larger specimens had proportionately lower ratios—that is, larger specimens were older—and predated the nuclear testing increase in carbon-14 in the ocean. Using a statistical model, the authors of this work were able to translate this ratio into an age for some 20 Greenland sharks, and found that the oldest ones had lived for 400 years (Nielson et al. 2016).

Eye lens tissue, once created in the early embryo, is metabolically stable. Once the lens is produced, there is no cellular turnover. The same analysis has been done on the lens in the human eye. (However, cataract surgery, replacement of the lens, can reset the age clock.) Of course, we have other means of establishing human birth dates.

How long does it take a *Nautilus* to grow a new chamber for its beautifully constructed coiled shell? *Nautilus* are from an ancient group, predating all of the other cephalopods, such as squid and octopus. They arrived in the fossil record 200 million years ago, and descend from the common ammonites, another group of shelled cephalopods. Knowing about the growth of the shell, a chamber at a time, and also how long it takes for a *Nautilus* to reach maturity aids conservation efforts for this increasingly exploited animal. Neil H. Landman (American Museum of Natural History), Ellen Druffel (UC Irvine, whom we met in chapter 7), Kirk Cochran (State University of New York at Stony Brook) and A. J. T. Jull (University of Arizona) did this work, taking advantage of the increase in carbon-14 as a consequence of nuclear testing in the 1950s and early 1960s.

First, the authors needed a separate the chronology for carbon-14 in the surface ocean, and this they got by sampling growth rings in coral from the Great Barrier Reef. The coral showed the span of carbon-14 during atomic testing, rising rapidly as carbon-14 entered the ocean, from 1959 until 1974, after which it leveled off and began to decline (see figure 4.6). Carbon-14 in the nautilus chamber showed similar changes. The authors concluded that their specimen hatched prior to 1957 with seven chambers and added chambers quickly over the succeeding year. Then the growth declined, with the last chamber taking much longer, seven months, to complete.

Understanding the turnover of heart muscle cells, called cardiomyocytes, might help with therapies in the recovery from heart attacks and cardiac tumors. However, concerns for safety have prevented studies in humans. Researchers in Sweden (Bergmann et al. 2009) achieved a workaround, using the bomb pulse from atmospheric testing (see chapter 4) to estimate the turnover of cardiomyocytes. Carbon-14 becomes part of the cells' DNA, and given the changing atmospheric concentrations of carbon-14 during the 1970s, '80s, and '90s, they were able to calculate how many of these heart muscle cells were renewed over a lifetime, and to inform potential therapies for heart disease. Research like this, referred to as "bomb-pulse research," is an emerging area of biomedicine, made possible by atmospheric testing of nuclear weapons.[7]

■ ■ ■

I will return later to the subject of carbon-14 dating, involving much older dates than the Shroud of Turin, Ötzi, or the Bronze Age. To get a perspective on these older, paleo, dates and how they are important to our current situation, we will have to "dive" into the ocean.

9

OCEAN CIRCULATION

Arnold Gordon looked with a combination of alarm and disgust at the small pen as it slowly inked its way toward deeper depths on the plotter paper (this was 1983), recording the conductivity (salinity) of the seawater. The ship had stopped for a "test" station at the beginning of an oceanographic expedition to make sure the conductivity-temperature-depth profiler (CTD) was working properly before heading hundreds of kilometers offshore, and beyond help.

Gordon and his crew sited the test station[1] off the coast of Cape Town, South Africa, a location called Cape Basin, in the southeast corner of the Atlantic Ocean, just north and west of the Cape of Good Hope. Looking at the trace, Arnold at first thought the pen on the plotter was stuck, and gave it a gentle rap. But the plotter pen recorded what it was told. It was responding. Well, then, maybe the sensors on the CTD were broken? The sensors consisted of a thermistor to record temperature and a conductivity cell from which to derive salinity. If they were not working, it would jeopardize the month-long cruise ahead of them. Fixing or recalibrating the sensors in those days required a company tech, not someone that is usually available in the ship's crew or science party. Replacing the sensors would mean returning to Cape Town, squandering expensive and unrecoverable ship days.

All these options ran through Gordon's mind in seconds. Then, his disposition went from disgust to astonishment. He recognized that the high conductivity numbers—high salinities—he was observing at depth in the Atlantic Ocean came from seawater that had originated in the neighboring Indian Ocean, far to the east, on the other side of Africa.

Further stations nearby confirmed the salinities the CTD recorded. They showed an anomalous water mass: a deep ocean eddy, a slowly spinning blob of water carrying higher temperatures and salinities than its surroundings. The eddy was quietly and unobtrusively making its way into the central South Atlantic.

Waters from the western Indian Ocean become part of the Agulhas Current, hugging the east African coast along South Africa and Mozambique. The Agulhas takes its name from Cape Agulhas, the southernmost tip of Africa. Agulhas is Portuguese for "needles." One story has it that the Portuguese, the best navigators in the early days of European exploration, noted that at this location, magnetic and geographic north were the same. The "needles" on their compasses aligned. Another story, more prosaic, has it that the Portuguese saw some needlelike rock formations along the coastline.

The Agulhas mirrors the North Atlantic's Gulf Stream, being a current that travels toward a pole along a western boundary of an ocean. The Gulf Stream is remarkable for its velocities, but the Agulhas moves with even higher energy. When it runs out of land to help steer it, the Agulhas becomes unstable and turns back on itself—retroflects—because it interacts with the Antarctic Circumpolar Current to its south, across the subtropical front in the Southern Ocean, flowing in the opposite direction, to the east. Now destabilized, some of the Agulhas water finds its way into the Atlantic as eddies pinched off from the main current, and the eddies travel westward into the Atlantic Ocean. Oceanographers nowadays call these eddies the "leakage" of the Agulhas (figure 9.1).

What surprised Arnold—and why he initially thought something was amiss—was the anomalously high salinity. If the eddy did indeed come from the Indian Ocean, it could represent the final link of something grand, an interconnected circulation throughout the global ocean. That extra *soupçon* of salinity contributes to a dynamic 8,000 kilometers away in the northern reaches of the Atlantic Ocean, where it all "begins."

■ ■ ■

If piecing together the circulation for the entire ocean less than 40 years ago seems like a recent discovery, it illustrates something about sampling 71 percent of the surface of the Earth, where no one lives. Oceanographic

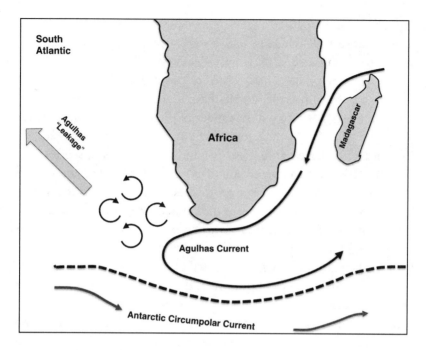

FIGURE 9.1

The Agulhas Current near Africa and in the South Atlantic Ocean. After the Agulhas passes the Cape of Good Hope at the bottom of Africa, it becomes destabilized and turns back on itself (retroflects) from its interaction with the Antarctic Circumpolar Current, staying north of the Subtropical Front (dashed line). Destabilized, the Agulhas then sheds and pinches off eddies (small circles with arrows), which travel into the South Atlantic as Agulhas "leakage."

research vessels ply the seas, and oceanographers confront the ocean with an attitude reminiscent of the seaman's prayer: "Oh God, thy sea is so great and my boat is so small."[2] At 120 million cubic miles (600 million cubic kilometers), figuring out how the ocean moves by measuring its physical properties, in what works out to be from teaspoon-size volumes, is tremendously hard—let alone what controls those movements, and then how they affect our climate. Some have said that if meteorologists sampled the atmosphere the way oceanographers sample the ocean, they'd be driving around the country in a Depression-era automobile and stopping every 50 miles or so to send up a kite. It was common, 30–40 years ago, for oceanographers to come off a ship and say, "Well, conditions were anomalous."

That's because there had not been enough observations to know the difference between "anomalous" and "usual," a symptom of the ocean, vast as it is, not being adequately sampled—not even close.

The sampling problem has receded of late. Nowadays, research vessels collecting those minuscule samples have become an endangered species. Much can be learned from observations of the surface ocean from Earth-orbiting satellites. As well, automated buoys, floats, and robotic "gliders" (sort of like undersea drones) search and sample the sea, sending data back to desk- and computer-bound oceanographers. For the present, however, one area seems immune to these sampling marvels. Various kinds of chemical substances are distributed unevenly throughout the surface ocean and with depth, and these can be used to unlock the ocean's puzzles, most helpfully by setting a time dimension for ocean circulation. Because they exist in sometimes vanishingly small quantities, oceanographers still resort to collecting seawater directly, up to hundreds of liters at a time. Sampling for these trace substances, or "tracers" can only be accomplished from ships, using large-volume water samplers, at a series of geographical locations, and over the entire two-and-a-half-mile depths (on average) of the ocean.

As I will show in the next chapter, understanding the distribution of carbon-14 and its use as a tracer means going to sea and gathering water samples from individual locations, not materially different than what mariners did 200 years ago. Since its discovery in 1940 and the recognition a few years later of its natural abundance, carbon-14 has played a central role in understanding the circulation in the global ocean. The distribution of carbon-14 in the ocean relative to the inorganic dissolved carbon (designated as "$\Delta^{14}C$," in parts per thousand, or "per mil") gives a time scale for the movement of deep ocean currents. When we analyze the skeletons of organisms left in ocean sediments, we can figure out what conditions were like in ancient oceans, and how those oceans affected climates in the distant past.

We can also follow carbon-14 through the ocean's food web and into the residues left from organisms' feeding and respiration. The significance of the ocean's carbon cycle to Earth's climate has only recently been appreciated. The findings in records of past climates (the subject of chapter 13) that atmospheric carbon dioxide and temperature change with respect to one another, and relatively rapidly, suggest that the ocean's carbon cycle participates in climate change. This relationship has stimulated research using carbon-14.

First, however, a little history.

■ ■ ■

One of the small pleasures doing shipboard work in the tropics—known as "barefoot oceanography"—happens when the CTD and its collection of attached water samplers arranged in a circle, called a "rosette," arrives back on deck after being sent down to the deep (figure 9.2).[3] It is hot work,

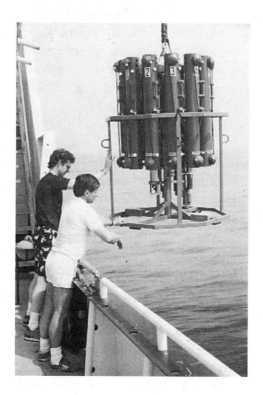

FIGURE 9.2

Rodolfo Iturriaga and Jim Luther launching a CTD rosette, 1987.

Photograph courtesy of John F. Marra.

the sun relentless. Once the water samples from the various depths have
been collected into bottles for analysis, those of us left on deck release the
water left in the samplers to splash on our toes or canvas shoes, sending
a chill up our legs and providing momentary relief from the heat. We are
mostly ignorant of the fact that our cooling-off replays the beginning of
deep-sea oceanography, two centuries earlier.

If there is such a category in civilization as a "humane slaver," then
Captain Henry Ellis can be so classified. After expeditions to the Arctic to
map out a possible Northwest Passage, he sailed three trans-Atlantic voyages on the slave ship *Earl of Halifax*, bringing captive Africans (captives
or slaves that African chieftains had won in battle) to Jamaica. Often slave
ships were becalmed near the African coast, and without wind, daytime
temperatures scorched the decks and turned interior spaces into ovens.
Ellis, who was also a Fellow of the Royal Society, was always looking for
novel ways of running his ships. For example, he installed "ventilators"—
essentially bellows, developed by the British clergyman and Ellis's friend,
Stephen Hales—to relieve the cramped and stifling conditions below
decks. Using the ventilators, he took good enough care of his shipboard
charges, crew and slave alike, that no one ever died during the transits to
the Caribbean.

Hales also gave Ellis a device he called a "bucket sea-gauge." The
sea-gauge was an enclosed wooden bucket containing a Fahrenheit
thermometer. The bucket had valves on top and bottom that would be
open only when the bucket, now weighted, was being lowered through
the water. In this way, water could be collected at a depth correspond-
ing to the amount of line Ellis's crew paid out. In those tropical waters
(approximately 25°N/25°W), to his astonishment and delight, it did not
take too much line to retrieve a sample of cold water underneath the
warm surface layer. Over some days, he was able to get six samples from
depths ranging from 360 feet to more than a mile beneath the surface,
and he found temperatures becoming constant at 53°F at the deeper
depths. [We now know the deep sea has temperatures near 40°F (~4°C);
Ellis's bucket warmed on its return to the surface.] Retrieving the water,
Ellis and his crew could have cold baths and chilled wine.

For Ellis, the discovery went no further than providing comforts for
himself and his crew. We know that because he related his findings by

letter to Hales in 1751: "This experiment, which seem'd at first but mere food for curiosity, became in the interim very useful to us. By its means we supplied our cold bath, and cooled our wines or water at pleasure, which is vastly agreeable in this burning climate."[4] But almost 50 years later, in 1797, an American living in Germany, Benjamin Thompson, heard about Ellis's observations and made the leap to propose a southward movement of polar water at depth, with a return flow back to the pole at the surface, reasoning, "It appears to be extremely difficult, if not quite impossible, to account for this degree of cold at the bottom of the sea in the torrid zone, on any other supposition than that . . . cold currents from the poles [flow in the deep]" (Brown 1968, 211).

Both Ellis and especially Thompson had kaleidoscopic lives. After his three trips as Captain of the *Earl of Halifax*, Ellis was appointed governor of Georgia (first, lieutenant governor, and later, royal), and was known for his good relations with the local Native Americans. He returned to the British Isles (he was born in Ireland), where he was sought after for his knowledge of the colonies, and the king of England then appointed him governor of Nova Scotia. He never took his seat, but a garrison was built in his name north of Halifax. Later he moved to Naples, engaging in further scientific research, and died there in 1804.

Thompson, born in Woburn, Massachusetts, moved at age 19 in 1772 to Rumford (the former name of Concord), New Hampshire, hired as a teacher. He married a local heiress, a propertied widow with government connections, a decade his senior. At the outset of the Revolutionary War, Thompson was a Loyalist, and came under suspicion by his neighbors of being a spy for the British. He was not a spy, but revolutionary feelings were strong enough (he was threatened with a tar-and-feathering) that he reckoned he had best clear out. Being unjustly accused and his status compromised, he realized he had no future in the colonies. He consequently offered his services to the Crown, and eventually headed across the Atlantic. In England, for his various services to British forces in support of the war, King George III knighted him. Thompson's work also got him elected as a Fellow of the Royal Society. He later moved to Bavaria and entered service to the Duke, who named him a count of the Holy Roman Empire in 1791. Thompson chose Rumford as the name for his title, maybe to retaliate for the treatment he got from his erstwhile neighbors, or maybe to honor his wife, whom he had left behind. In Bavaria, Count Rumford kept busy. He came up with the idea of using dried bread pieces to thicken soldiers' soup—the crouton. He encouraged potato cultivation (again, for the military), improved a stove, and invented a percolating coffee maker, among other innovations. He originated soup kitchens for the poor and created a large urban park in Munich. It was during this time that he developed his ideas about ocean circulation. In the early nineteenth century, he spent time in both Paris and London; he died in Paris in 1814.[5]

Another 50 years after Count Rumford hypothesized an oceanwide circulation from Ellis's observations, Emil Lenz, finding cold water at shallower depths near the equator compared to more temperate regions, diagrammed two circulation cells over ocean depths—one to the north of the equator, and its mirror to the south (figure 9.3). What we know today has changed more in detail than in conception.

The bold ideas of two centuries ago seemed unmatched by the meager data that supported them. The intervening years have filled in the observational gaps, from voyages such as the British *Challenger* expedition in the 1870s and the *Meteor* expedition mounted by Germany in the 1920s. Both changed ocean science decisively, but discoveries about the physical

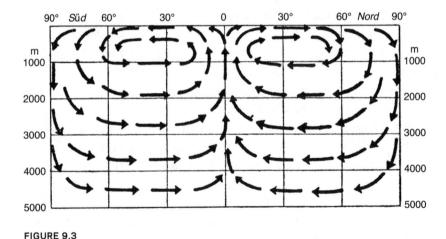

FIGURE 9.3

Emil Lenz's hypothesized deep ocean circulation, from 1831.

oceanography of the deep sea were a by-product. Deep-sea hydrography mattered less than, for the *Challenger* expedition, testing the hypothesis of a lifeless abyss or, for the *Meteor* expedition, extracting gold from sea-water to pay Germany's reparations from World War I. A lifeless abyss was decisively disproved at the *Challenger*'s first deep-water station. Using bottom dredges, the crew brought abundant life up from the bottom of the ocean everywhere they sampled. Both expeditions produced much hydrographic data on the deep sea, but did not advance any ideas about the abyssal circulation beyond that of Rumford and Lenz.

■ ■ ■

At most drinking establishments and pubs (but not Irish pubs), you can order a "Black and Tan."[6] If you can also watch the bartender, he will half-fill a pint glass with a pale ale and then very carefully pour or spoon in stout, so that the stout floats on top. A Black and Tan, so prepared, consists of the (black) stout on top of (tan) ale, with the stout's foam making a drink with three layers. Differences in the density of the two brews (and the head) allow the layers to stay that way, with the less dense portions (the foam and stout) on top of the heavier (ale). Even though it's called stout, and dark, it's actually less weighty than the pale ale.

We can think of the ocean in a similar, maybe more sober, way. The surface layer of the ocean (the foam) is lighter, less dense by virtue of being warmed by the atmosphere and, ultimately, the sun. The ocean bottom layer is the heaviest and, in most locations, far removed from the effects of the sun and wind. There is also a middle layer in the ocean (the stout in the Black and Tan), the thermocline. The thermocline connects the warm, less dense surface layer and the cold, dense, deep layers, with temperature rapidly changing from warm to cold going from shallow to deep. The two contributors to the density of the layers are temperature and salinity, although in most areas, temperature exerts the greater control. Salinity's part in the ocean circulation is like the "Fifth Business" part in an opera or drama: a minor role, but crucial to the plot.

Aside from jiggling the glass in our Black and Tan, how can the stout and ale layers be disrupted or mixed? In the ocean, it can be cold temperatures or high salinity, or better, both. Colder temperatures at the surface increase the water's density, as does a higher salinity, and heavier water will sink and disrupt the layers. If a warm ocean gives up its heat to a cold atmosphere, the ocean gets colder, and also becomes saltier, together making the surface ocean denser. The big drivers are the winds and the sun. The wind imparts its energy to the water and moves the surface ocean around, and penetrating the surface layers, disrupts them. The sun has the opposite effect, warming the surface and stabilizing the layering of the ocean. The interaction of wind and sun, played out over the scale of ocean basins, drives the global circulation, with the help of Earth's spin.

Historically, oceanographers have separated shallow wind-driven ocean movements from the circulation in the abyss. In the 1950s, Henry (Hank) Stommel, at the Woods Hole Oceanographic Institution, published two landmark (or maybe better, watershed) works, the first explaining the surface circulation in ocean basins, and the second, a theoretical underpinning for the global, deep-ocean circulation (Stommel 1958; Stommel and Arons 1960). He explained the existence of the Gulf Stream in the Atlantic and, by extension, the other so-called Western Boundary Currents, such as the Kuroshio, passing Taiwan and Japan, and the Agulhas, off the southeast coast of Africa, that Arnold Gordon studied. For his second contribution, the circulation in the ocean's abyss, Stommel proposed a deep current originating near the southern tip of Greenland and flowing south, following the contours of North and South America in the Atlantic. That second bit of work showed particular audacity. No one had ever observed current flows in the abyss.

Stommel's latter work described what came to be known as the thermohaline circulation (THC), "thermo-" indicating temperature and "-haline" indicating salt, or salinity.[7] Temperature and salinity give seawater its density, how much it weighs per unit of volume. Differences in density drive the thermohaline circulation through the deep ocean. But it is the circulation driven by the wind that sets up the conditions that allow the density differences to operate. You need the wind to drive the surface currents to a place where their increased density can disrupt any layering and cause surface water to sink, or mix downward, into the deep sea, thereby becoming part of the thermohaline circulation in the ocean's depths.

The surface and deep ocean circulations are not independent, although they do operate on different timescales, and vastly different volumes of ocean. The deep ocean circulation moves about 90 percent of the ocean's volume, but over a much, much longer time. Nowadays the term thermohaline circulation (THC) has been replaced by the meridional overturning circulation (MOC), and in the Atlantic, the AMOC. "Meridional" refers to the fact that the axis of the overturning is north-south, along longitudinal lines of meridian. AMOC might be pronounced "amok" (as in "run . . ."), the only Indonesian word adopted into English. "Amok" might be a better descriptor, or at least an expressive homonym. Unlike the THC, the MOC does not indicate a driver. Oceanographers recognize that a density difference alone will not drive a circulation; as we will see, winds at the surface and mixing in the deep sea are probably more important.

Looking at the global ocean, where do we find areas where the ocean's layers, the Black and Tan, can be disrupted? Where, at the surface, is cold water combined with high salinity to produce the necessary high density? For temperature, the polar regions are cold, obviously, because the sun never gets very high in the sky, and they receive much less solar radiation. What about salinity? Over the global ocean, the North Pacific has the lowest salinities, and these are maintained because any evaporation from the ocean, gets trapped by the North American continent. The prevailing winds at those latitudes, carrying water vapor from the Pacific, generally get no farther than the American cordillera, the Rocky Mountains. Keeping the wet atmosphere west of the Rockies contributes to the low salinity in the North Pacific (and a rainy Pacific Northwest). Contrast that to the Atlantic, where water vapor evaporated into the atmosphere has no mountains to get over, and is sent to Siberia.

These imbalances have themselves to be balanced, of course. The North Pacific has not become increasingly fresh, and the North Atlantic has not become even saltier. I will now attempt to explain how the global ocean circulation maintains the conditions observed. Carbon-14, in its various tags, plays a central role. It is a clock contained in every water mass, not to say every water particle, keeping track of the age of the water.

I'll use the North Atlantic as an example, or paradigm, for what goes on (with some variation) in the other oceans. Compared to the Pacific, which occupies nearly a whole hemisphere of the Earth, the Atlantic is small. A famous oceanographer of the late twentieth century, John Martin, who worked almost exclusively in the Pacific, referred to the North Atlantic as "that cute little ocean."[8] "Cute" can be a virtue for scientific understanding, and most of that understanding of the ocean has come by way of research in the North Atlantic, surrounded as it is by the origins of oceanography in Europe and successors in North America. Figure 9.4 shows the North Atlantic Ocean; the surface, or wind-driven, currents; and the locations where surface waters can descend to the deep ocean. How did this distribution of currents arise?

The wind-driven circulation, as the name implies, happens in the surface ocean. Picturing the North Atlantic Ocean, we can see how temperature and salinity, influenced by the atmosphere and sun, move surface ocean water around, sort of like stirring water in a bathtub. The first factor is the heat distribution on Earth. The tropics, between about 23 degrees latitude north of the equator (Tropic of Cancer) and a little more than 23 degrees

FIGURE 9.4

North Atlantic Ocean circulation. The surface circulation is in gray, the deep circulation in a darker shade of gray. Dashed arrows indicate the approximate locations of the northeast trade winds (from North Africa and the Mediterranean) and the westerlies (over North America). Deep water is formed near Greenland, east of Iceland, and in the Labrador Sea.

south (Tropic of Capricorn), receive the most direct solar rays, all year around. The midday sun sits almost directly overhead in all months, moving a little bit seasonally, as the Earth, tilted by about 23 degrees, orbits the sun. The tropical ocean heats up. A warmer ocean means the water is less dense, and therefore relative sea levels rise slightly higher. Gravity drives the higher water downslope, to the north and south. There we have the first component of motion: a heated ocean in the tropics causing a flow to the north and to the south.

The next factor is the prevailing winds. In the north tropical latitudes in the Atlantic are the northeast trade winds, occupying a band of latitude from the equator to about a latitude of 25 degrees. A weird convention, maybe because of atmospheric scientists and oceanographers not talking to each other, is that winds are named for where they are *from* and ocean currents are named for the direction they flow *toward*. So, the northeast trade winds are from the northeast, from Europe and Africa. The trade winds get their name from early European navigators sailing to the New World with the wind behind them, to bring back America's riches. These northeast trade winds drive a westward ocean current. Further north are winds originating in the west, the westerlies, which drive the ocean toward Europe, to the east from North America.

Now we have elevation and the flow of heat in the ocean driving a flow to the north, a north component to the circulation. Then we have the winds, in the tropics pushing the surface ocean toward the west, and to the north, in what is called the temperate zone, pushing the ocean to the east. The last major factor is the rotation of the earth itself, called the Coriolis effect.

Gaspard-Gustave de Coriolis (1792–1843) was a French mathematician and scientist who became interested in how water moves in industrial applications, at a time when industry was powered by waterwheels. His research is an example of how seemingly unconnected phenomena are shown to be connected after all. Coriolis found that an additional "force" exists on rotating surfaces, and although small, it was considered important to the industry of the time. He never thought his work applied to anything larger than waterwheels, let alone the Earth as a whole, but later in the nineteenth century, meteorologists began realizing that the same effect could be used to understand motions in the atmosphere, and later, by ocean scientists, motions in the ocean. The Coriolis effect is small, but it becomes significant when operating over large distances, like continents or ocean basins, and over longer periods of time. We, dwellers of the land, are all attached to the Earth by gravity, so where the Earth goes, we go. But the atmosphere and ocean, being fluids, move more independently from the solid Earth. And that's where the Coriolis effect becomes noticeable.

One useful way of seeing this is to imagine a plane trip. Sitting in your seat on the plane, you now move independently of the motion of the Earth. Say you take off from Bogotá, Colombia, and head to New York City. While you, in the plane, fly at something like 30,000 feet, the Earth continues to rotate beneath you. If the pilot does not compensate for that, you might end up landing in Pittsburgh. New York has "moved" to the right compared to the airliner in the sky; the frames of reference are different. In the northern hemisphere, the Coriolis effect deflects air or water to the right. In the southern hemisphere, the effect is to the left of the direction of motion. Because it is dependent on rotational velocity, the Coriolis effect changes with latitude, becoming much stronger at the poles and vanishing at the equator. It simply takes longer to make one spin of the Earth at the equator than at the north or south pole.

Together, the flow of heat in the tropics by gravity, the influence of the winds, and the Coriolis effect create the North Atlantic Gyre, a clockwise circulation caused by waters from the tropics moving north, being deflected to the right (eastward) by the trade winds and, much further north, to the right again by the westerlies (southward). The return flow, from the north to the south, in the eastern half of the North Atlantic, deflects to the west (to the right again), completing the circle.

One important permutation is also caused by the rotation of the Earth along with the winds. Fluids, whether in the atmosphere or water, tend naturally to rotate. In the atmosphere, we see this rotation in, for example, hurricanes and tornadoes. In the ocean, we see the rotation as whirlpools, or eddies. Under the influence of the wind, in the North Atlantic, the whirlpools will rotate clockwise. Physical oceanographers call the tendency to rotate "vorticity." The other kind of vorticity, planetary, comes from the rotation of the Earth, and is in an opposite direction from that caused by the wind. Physical oceanographers have derived a relationship that shows that vorticity, whether from the rotation of the Earth or from the influence of the wind, is conserved. Like the Coriolis effect, planetary vorticity depends on latitude, declining from the poles to the equator. The way the conservation of vorticity plays out over the North Atlantic Ocean produces the currents we see at the surface.

Stommel knew from observations that the ocean in the eastern North Atlantic flows slowly to the south, explained because planetary vorticity from the rotation of the Earth declines with latitude. That means that, to conserve vorticity, the wind-driven vorticity must decline as well. To complete the gyre circulation, the ocean on the other, the western, side of the Atlantic, flows to the north. However, here, in terms of vorticity, the opposite happens. The two different sources of vorticity, wind and the planet, reinforce one another as the ocean water moves north. Ocean water, still rotating, will also feel friction from the ocean surface, from the bottom, and from coastlines, and friction will slow the rotation down. But since vorticity has to be conserved, friction must increase as well, to balance the increase in vorticity as the water moves north. The only way to increase friction is to increase the speed of the water, and that produces the Gulf Stream. We like to think of it as a "stream," but a better metaphor is a "wall" of water speeding to the north.

The Gulf Stream, a warm, salty current coming out of the Caribbean and the Gulf of Mexico, was first identified by Benjamin Franklin in colonial Philadelphia from the shorter times his mail took to arrive in England compared to mail sent in the other direction, an early example of armchair oceanography. Ship captains had long taken advantage of the Gulf Stream to give them an extra couple of knots' push toward Europe. The South Atlantic mirrors the North with a counterclockwise gyre, and the Brazil Current, flowing toward the Antarctic, is the South Atlantic's western boundary current.

The Gulf Stream travels north and east up the coast of North America, ending up south of Newfoundland, where it becomes the North Atlantic Current. Cold weather, and cold air coming off of Canada, blows over this warm water and extracts its heat. As a result, the North Atlantic Current cools, while still having a high salinity. The cold, salty water near Greenland and Iceland is some of the densest seawater in the global ocean, and has nowhere to go but down to the abyss. There, it becomes what is known as North Atlantic Deep Water (NADW). What happens after the cold, salty water descends a mile or so beneath the surface is complex, but the NADW, from its more or less distinct combination of temperature and salinity, its "T-S" character, can be followed south and into the South Atlantic, into the Southern Ocean, west across the Indian Ocean, and eventually into the North Pacific. There, it rises to mid-depths, the thermocline, and returns, through the Indonesian archipelago, across the Indian Ocean, and back to the Agulhas.

What Arnold Gordon put together in the southeast Atlantic, the Cape Basin, was the return path for the deep circulation from the Pacific, the final piece to the puzzle. The Agulhas eddies cross the South Atlantic, adding salinity, and make their way north. They mix with water from the North Atlantic Subtropical Gyre and eventually become part of the warm and salty Gulf Stream. The Gulf Stream brings that warm, salty water to the northern reaches of the North Atlantic where, giving up its heat, it becomes cold and salty and forms the North Atlantic Deep Water, completing the cycle.

Gordon identified a couple of key locations along the global path, one of which was the Agulhas, leaking its water properties by way of eddies, and the point of entry into the Atlantic Ocean from the Indian. He surmised that the other key location was in Indonesia, the only low-latitude boundary between two oceans, the Pacific and the Indian. The "Indonesian Throughflow" connects the tropical Pacific with the likewise tropical Indian Ocean. Gordon (1986) called this worldwide scheme a "global circulation cell," the fundamental part of the ocean's deep circulation. And it all came about from those strange observations in the Cape Basin.

A year later, Wally Broecker (1987), also at Lamont, and with a knack for the apt metaphor, called the circulation the "global ocean conveyor belt." The ocean conveyor belt expresses the idea of how the ocean can transport properties across the seas and back again. Some of these properties are passive tracers, riding on the conveyor belt, maybe changing along the way, increasing or decreasing depending on how the ocean mixes, or are otherwise transformed by organisms or simple chemistry. The conveyor idea also implies a closed loop: water that sinks in the North Atlantic later returns.

Earlier, I mentioned the heat and salt balance. The Atlantic is not getting saltier, and the Pacific is not getting continually fresher. The conveyor idea furnishes a solution. It carries the salt surplus from the Atlantic (while sending water vapor to Europe and Asia) to balance the salinity in the Pacific, which is subject to excess precipitation.

The conveyor belt is a gross oversimplification. Criticisms of the conveyor are legion, targeting it because it is oversimplified. But those criticisms miss the point. The conveyor is notional, or conceptual— a hypothetical construct. The idea of the conveyor belt is very useful, showing how ocean properties pass through the ocean, and what might influence the circulation. It is a loop, which in its conception, drove ocean science forward.

Still, no one was sure how long it took for Broecker's global conveyor belt or Gordon's global circulation cell to complete the loop. Depending on your point of view, carbon-14 either simplifies or complicates the overall picture. But importantly, carbon-14 is about the only way to put "time" into the global ocean circulation. In the next chapter, I'll describe how carbon-14 is distributed in the ocean, and how it gives us a measure of time.

10

CARBON-14 IN THE OCEAN

Henry Stommel theorized deep ocean circulation pretty much only in terms of which way the deep-sea currents would flow and where they would be strongest. Ingenious though it was, the scheme did not have a useful timescale other than "slow." Or "very slow." The distribution of North Atlantic Deep Water (NADW) was identified from its characteristic temperature and salinity, but that combination indicated little as to the transport of water, at least at the global scale.

How fast (or, in this case, slowly) does the water move? What is the volume transport? How long does the water stay in one ocean basin over another? How long does it take to complete the global circulation cell, or one loop of the global conveyor? Answers to these questions give information on the timescales of climate change, how the ocean mixes, and how the fertility of surface waters are revived in terms of nutrients, chemicals reconstituted from organic matter that has sunk to the deep. Stommel, however, influenced what came next in deep-sea measurements to answer these questions: neutral-density, or "Swallow," floats and large-volume water samplers (see Warren and Wunsch 1981).

In the 1950s and 1960s—still early days for oceanography and marine geophysics—answering questions about the ocean usually involved solving a technical problem, improving the technology, and devising a new kind of measurement or sampling platform. Oceanography has been doubly hindered because there has not been much of a market for sampling gear or for analytical equipment designed for the rigors experienced on ocean voyages. There was no such thing as an "off-the-shelf" item; there was no "oceanography store."[1] There was one company—General Oceanics in

Miami, Florida, established in 1966—that supplied oceanographers with water samplers for routine hydrographic measurements. The problems of capturing water at depth and recording an accurate temperature at that depth had been solved decades before.[2] Pretty much anything beyond that required creativity.

John Swallow, at about the same time and independently of Stommel's research, was developing a subsurface drifter in England that could be carried along the slight currents in the deep sea. These "Swallow floats," as they came to be known, were designed to match exactly the density of a water mass, making them neutrally buoyant—neither floating up nor sinking below a prescribed density horizon. Considering that the density of seawater changes only slightly, despite driving massive currents, targeting a specific depth or density layer in the ocean makes for a technical challenge. The difference between the density of North Atlantic Deep Water and its surroundings might be from 1.0267 grams per cubic centimeter to 1.0234, a difference of only 2 percent.

The floats were otherwise simple, with two 10-foot lengths of aluminum tubing Swallow got from building construction scaffolding, mated together. The top tube was the neutrally buoyant float, and the bottom tube contained batteries and a sound transducer emitting a coded signal. The floats were launched or thrown from a ship, found their prescribed depth, and drifted on their own. Their position could be fixed by using a hydrophone, a sound receiver, on a surface ship. By moving the ship with respect to the sound, or using two ships each with a hydrophone, the float's position could be triangulated and then tracked over weeks or months. From the change in position with respect to time, a net velocity could be calculated.

After some early technical disappointments, a joint British-American expedition was formed, consisting of the British ship *Discovery II* and Woods Hole Oceanographic Institution's *Atlantis*. The ships left Charleston, South Carolina, in the spring of 1957 and deployed eight floats just east of the Gulf Stream. In this particular case, "gone south" referred to a direction and success, and seven of the floats did just that. The floats corroborated Stommel's deep western boundary current, a major achievement in verifying a theory through observation.

Swallow floats were good, but similar to the surface ocean, the deep sea was found to be dominated by eddies—not nearly as energetic as those at the surface, but enough to confuse the movement of the floats. It proved impossible to establish any further details of the deep ocean circulation beyond the barely identifiable deep current on the western boundary of the North Atlantic. For subsequent observations, the float trajectories muddled more than enlightened. They might end up going in the opposite direction, or just not move at all. The floats simply could not be deployed long enough to average over the deep ocean's shorter-term and seemingly erratic variability. They could only be tracked for months, while the global circulation had to be measured in decades.

Stommel recognized that to advance understanding of the deep ocean required something in its chemistry, something that could identify the very slow deep ocean flows in spite of the eddies. Also, the chemical component had to contain a clock. Carbon-14 was just that sort of chemical species, and a tracer of ocean movement.

Barring outside influences, carbon-14 atoms decay away with a rate constant of about 8,300 years. That means that, on average, a carbon-14 atom survives for about 8,300 years after it is created in the upper atmosphere.[3] (The half-life, as noted previously, is ~5,730 years.) A lifetime of 8,300 years is a good length of time for figuring out the age of the waters in the deep ocean. It is long enough to be able to average over much of the ocean's eddy variability, but not too long to be able to detect changes to ocean circulation.

Waters near the surface will exchange their properties with the atmosphere, including CO_2, so these waters will contain the highest amount of carbon-14 (in CO_2), and will therefore be the youngest in the ocean. Once that water sinks, it loses contact with the surface. With no further air-sea exchange, the water mass "dies," so to speak, and the radioactive clock begins ticking, the same clock used in carbon-14 dating.

As I discussed in chapter 9, the warm, salty waters of the Gulf Stream and North Atlantic current arrive in the far North Atlantic, with its Arctic weather. The warm waters give up their heat to the cold atmosphere, creating a very dense water mass: water just a few degrees Celsius above freezing, and salinities at about 35 parts per 1,000. At two spots in particular—one just south of Greenland, and another area near Iceland— the water has a high enough density that it can mix with water from the deep ocean, or descend to depth. These small areas of the ocean directly connect the atmosphere with the deep sea. The descending water mass, the North Atlantic Deep Water, has a signature characteristic in its salt content and temperature. Those are not enough to help with the time component of the circulation, but the NADW also has a signature in carbon-14 from the atmosphere. Once the NADW loses its connection to the surface, the carbon-14 becomes a timekeeper.

It was not too long after the discovery of carbon-14 in the 1940s that oceanographers began looking for it in the ocean. Some of the earliest measurements date back to 1950 and were done at Lamont (Broecker et al. 1960). In those days, Lamont's ship, the RV *Vema*, traversed the ocean; every day at noon, the crew would take a core of ocean sediment, take a bottom photograph, and tow a net for plankton.[4] When the opportunity arose, they would sample the deep ocean using a Gerard barrel.

Robert ("Sam") Gerard, educated as a geographer, lived with his wife and family down the hill from the Lamont Geological Observatory (as it was called then) and occasionally hung out at Lamont's geoscience library in Lamont Hall. Lamont Hall was also where Maurice Ewing, Lamont's first director, had his office. Sam became excited by the science at Lamont, and in 1954 Ewing invited him to join the science party aboard the *Vema*. In those years, the Atomic Energy Commission (now the Department of Energy) looked for ways to dispose of waste from nuclear power plants, and the deep sea was an obvious candidate. After all, no one (at least no human) lived there, and the environmental impact was thought to be minimal. Evaluating the deep-sea option for waste disposal meant sampling at great depths for volumes of seawater that were at the time unheard of. Sam, with a talent for things mechanical, solved this problem, producing a sampling device that bears his name: the "Gerard barrel," or "Gerard sampler."

The sampler (figure 10.1) was designed both to capture a large volume of seawater and to flush itself while descending to a prescribed depth, attached to a wire rope. A lead weight designed to travel along the wire—called, appropriately, a "messenger"—was sent down from the surface. Arriving at the sampler, the messenger tripped a valve, closing it and capturing a volume of water. The sampler had additional means of pinpointing its actual depth, to account for the angle of the wire to which it was attached, coming from the ship's deck winch, and for the motion of the ship itself against the seas.[5]

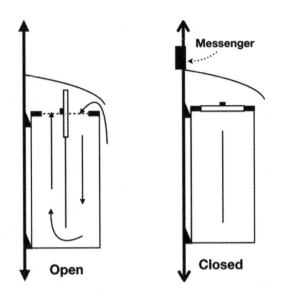

FIGURE 10.1

The Gerard sampler. On the left, the sampler is in the open position; on the right, it is closed, triggered by a "messenger." The arrows indicate how the sampler is flushed on its way to the deep, and the partition forces water to the bottom while it is being lowered. (Like the early oceanographic samplers, the Gerard sampler incorporated two thermometers that flipped vertically when the messenger closed the sampler's valves. See note 2 of this chapter.)

A seawater sample of 250 liters supplied enough dissolved carbon to assay for carbon-14. In the *Vema*'s laboratory, the sample had to be acidified to turn all the bicarbonate and carbonate dissolved in the water to CO_2, and the CO_2 captured and then oxidized to a solid, to be stored and returned to the laboratory at Lamont. These are the same procedures Kamen and Ruben used when they first identified carbon-14. With 250 liters of water, Wally Broecker, who headed up Lamont's carbon-14 analyses, could easily assay for carbon-14.

If carbon-14 can be used as a clock to track the ages of waters in the deep sea, and if Stommel's ideas are correct, then samples taken at various locations should conform to those ideas. Carbon-14 in the deep sea should decrease going south from the areas where NADW was formed, and in the western North Atlantic where Stommel hypothesized a south-flowing deep western boundary current. Carbon-14 should decline further in the South Atlantic, and decline even further into the Indian Ocean, finally recording the lowest values in the North Pacific. How did carbon-14 do?

The samples of oceanic carbon-14 in the late 1950s were limited in number, and also limited geographically to the Atlantic Ocean. But the values obtained showed, roughly, the expected decrease in carbon-14/carbon-12 moving from the far north to the south (Broecker et al. 1960). The numbers validated, at least somewhat, the ideas of the age of the deep ocean waters. In the Atlantic, the youngest deep water—that is, the water that has most recently experienced the atmosphere—was found to be along the east coasts of North and South America. The part of the Atlantic nearer Africa and Europe was much older.

In the early 1960s, after his collaboration with Swallow's floats, Stommel met with Broecker at Woods Hole and suggested a more extensive survey for carbon-14 in the ocean. He also pointed out that the National Science Foundation was looking for projects to support as part of the International Decade of Ocean Exploration (IDOE).[6] Broecker saw the opportunity and enlisted Derek Spencer, a colleague of Stommel's at Woods Hole, and Harmon Craig at Scripps. Arnold Bainbridge, also at Scripps, was proposed as the project director. Thus was born a program for the first extensive survey of the ocean's geochemical properties: GEOSECS.[7]

GEOSECS, a rough acronym for the Geochemical Ocean Sections Program, began in the late 1960s, to plan and schedule the cruises aboard research vessels and to establish procedures and methods. The overall goal of the program was to get a geographic picture of the dissolved chemical species and radioisotopes in the ocean by sampling pole to pole, from the Arctic to the Antarctic, in each ocean. GEOSECS was the first program to sample all the oceans over their entire depths in a systematic manner for their chemical properties.[8] The scientists on the program measured, among other things, the phytoplankton nutrients nitrate, phosphate, and silicate; oxygen dissolved in the water; and dissolved non- or inorganic carbon, as well as the isotopes of carbon, carbon-13 and carbon-14. Radium-226 was also assayed, as were, of course, temperature and salinity. GEOSECS happened before the advent of accelerator mass spectrometry (see chapter 7), meaning that large volumes of water had to be captured, and for that they used the Gerard barrel. The version of the Gerard barrel used on GEOSECS for isotope analyses was made of stainless steel and fabricated in Germany. Broecker noted that each cost the equivalent of a Mercedes-Benz.[9]

The proportion of carbon-14 to the total amount of dissolved inorganic carbon (as carbon-12) indicates the age of a water sample—how long it has been away from the surface and exchanging with the atmosphere. Comparing carbon-14 to the total inorganic carbon-12 normalizes it against the natural variability of the total amount of nonorganic carbon present. The GEOSECS program found that the deep ocean in the northern North Atlantic has the highest proportion of carbon-14 to total dissolved inorganic carbon (DIC). That proportion decreases in samples further south, into the South Atlantic Ocean, the Southern Ocean near Antarctica, and the Indian Ocean. The North Pacific samples contain the least amount of carbon-14. In figure 10.2, Minze Stuiver, Paul Quay, and H. G. Ostlund

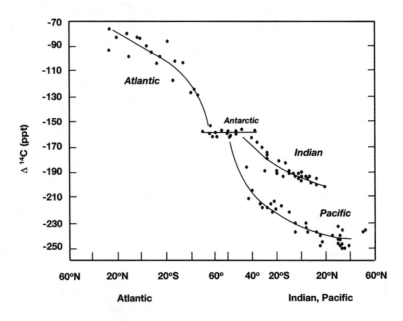

FIGURE 10.2

The distribution of carbon-14 relative to carbon-12 in carbon dioxide ($\Delta^{14}C$, parts per thousand, ppt) in seawater from various locations in the global ocean (after Stuiver et al. 1983). The "Δ" notation sets the values in a useful range.

plot the ratio between carbon-14 and total inorganic carbon, and in this way estimate how long the water has been away from exchange with the atmosphere, or how long it has been since the water near Greenland and Iceland descended to the deep. GEOSECS documents the decline through the Atlantic, and further declines in the Indian Ocean and the Pacific. Stable concentrations occur around Antarctica. Antarctica touches all three oceans, and the surrounding waters wheel around the continent.

So far, the carbon-14 data support the nascent ideas about global circulation, later figured to be the "global circulation cell" and the "global conveyor." A global map of the distribution of carbon-14 in the ocean is shown in figure 10.3.

The GEOSECS data indicate that if the declines in carbon-14 relative to total inorganic carbon from the Atlantic to the Pacific were caused entirely by radioactive decay, the deep waters would arrive in the North Pacific about 1,700 years after they exited the surface layer in the North Atlantic. In this, the simplest explanation, GEOSECS achieved a highly significant finding about the ocean's global circulation: its timescale. Otherwise, GEOSECS did not reveal any big surprises about the ocean, although the carbon-14 distributions greatly helped delineate ages and mixing in the deep sea. More importantly, as the first comprehensive survey of ocean properties, GEOSECS provided oceanographers with the data to develop ideas about how the ocean worked. Those data generated a multitude of scientific contributions over the years, and are still consulted and used today. GEOSECS led to later programs with similar objectives.[10]

The immediate data from GEOSECS can only be taken so far. Learning how to use carbon-14 as a tracer for ocean circulation and the ocean's carbon cycle is just as tangled and complex as the results it provides. Ocean mixing (the reservoir effect), variations in where the deep water in the ocean comes from, deep circulation variabilities, carbon-14 produced by atmospheric testing of nuclear weapons, the Suess effect, and, of course, biology all complicate the picture. I consider these complications in turn.

■ ■ ■

When carbon-14 is assayed in the dissolved inorganic carbon anywhere in the surface ocean, we find a surprising result: the surface water is "old"

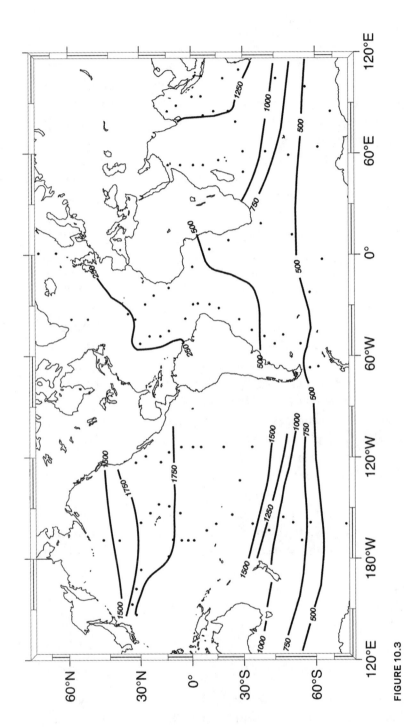

FIGURE 10.3

The difference between carbon-14/carbon-12 at 3,000 meters and the same value at the surface, expressed as an "age" of the deep ocean water. Waters in the North Atlantic clock in at about 250 years old; ages in the North Pacific are greater than 1,500 years (after Broecker et al. 1984).

by hundreds of years. If the surface ocean continuously exchanges with the atmosphere right above it, how can this be?

It turns out that the exchange—the mixing—with deeper ocean waters exercises much greater influence on surface-water dissolved carbon than does exchange with the atmosphere. Carbon in the atmosphere enters the ocean at a rate that is about 10 times slower than the rate at which the surface ocean mixes with the deep ocean. The slow exchange at the ocean surface results from carbon dioxide reacting with water to produce carbonic acid and carbonate, which I discuss below. The deep ocean, isolated from the atmosphere, will show greater decay of carbon-14 and older ages, and to the extent that that water mixes vertically, the surface ocean will also be old. If the surface of the ocean is covered by ice, it cannot exchange with the atmosphere, and the surface water will appear even older, again because greater mixing from the deeper waters will "age" it.

The differences from the expected values of carbon-14/carbon-12 can range from 100 years in the North Atlantic to 700 years in the Pacific. Where mixing between surface and deep ocean waters is more intense, such as in the polar oceans, the surface waters can be as much as 1,400 years old, very much depleted in carbon-14. In the central gyres of the ocean—for example, near Bermuda in the Atlantic or Hawaii in the North Pacific—the waters are more stable, there is less vertical mixing, and the atmosphere exerts a greater influence. Where waters upwell from depth, such as off the coast of Peru and California, and at the equator in the Pacific, the ages in the surface water reflect the deeper ocean to a greater extent. These anomalous ages are communicated to fish. Catch a fish in the surface ocean and analyze it for its carbon-14, and the fish would seem to be very much older than its actual years, and older than a comparable land vertebrate, one of the large cats or a wolf.

In chapter 9, I focused on the North Atlantic and North Atlantic Deep Water formation as a paradigm, a way of simplifying the explanation of ocean circulation across the globe. However, deep ocean water forms in an additional, different region of the ocean, further complicating age-dating with carbon-14.

The Antarctic Circumpolar Current, as its name implies, cycles around the Antarctic continent in a clockwise fashion—that is, from west to east.

As I discussed in chapter 9, the rotation of the Earth (the Coriolis effect) causes the waters of the Southern Ocean to turn to the north. The northward deflection, the waters moving away from Antarctica, produces a generalized upwelling around the Antarctic continent. Even so, the so-called Antarctic Bottom Water (AABW) forms in the Antarctic's coastal seas, such as the Weddell Sea and the Ross Sea. While water nearer the surface upwells and mixes over depth, AABW, which is even denser than North Atlantic Deep Water, creeps just above the ocean bottom as it travels north. AABW contributes much less to the deep-ocean circulation than its counterpart in the North Atlantic, but it is older. The waters of the North Atlantic, sinking to form North Atlantic Deep Water, have previously spent considerable time at the surface and have had a longer time to exchange with the atmosphere, increasing their carbon-14 inventory. The upwelled waters around Antarctica, the source of AABW, are already older, and more deficient in carbon-14. These variations in source waters make the age of the deep water of the Atlantic a more complicated mixture of ages than the global circulation scheme might imply.

Less significant complications to the age of the deep ocean come from the aforementioned within-basin variations, microbial respiration in the deep sea, and the carbon-14 produced during weapons testing. The east-west variations in carbon-14/carbon-12 in the North Atlantic suggest a deep gyre circulation that runs counterclockwise, counter to the gyre at the surface. The deep North Atlantic gyre is predicted by Stommel's theory.

■ ■ ■

As mentioned in chapters 4 and 7, in the 1950s, during what some call the height of the Cold War, the United States and the Soviet Union carried out atmospheric testing of their nuclear weapons. Detonation of these devices produced a whole host of radioactive isotopes, including strontium-90 and helium-3 (tritium), as well as carbon-14. Fortunately, after years of negotiation, a test-ban treaty was concluded in 1963, halting atmospheric testing (and moving it underground). For a time, then, there was an atmospheric spike, doubling the quantity of carbon-14 above that produced naturally, cosmogenically.

In chemistry and biology, there is an experiment called "pulse-chase." In this type of experiment, the subject is exposed to a labeled (usually radioactive) substance for a period of time (the pulse). Then the experimenter removes the pulse substance, stopping its administration or replacing it with an identical unlabeled substance (the chase). The pulse substance can then be traced through the system. In this case, bomb-produced carbon-14, usually called "bomb radiocarbon," is the pulse, and after the cessation of atmospheric testing, that time period becomes the chase. Bomb radiocarbon was introduced into the atmosphere, and after a time, it disappeared from the atmosphere. After subtracting the natural radiocarbon, it is possible to follow the bolus of newly radioactive water as it sinks and spreads through the North Atlantic. Instead of age-dating the deep ocean, finding carbon-14 from nuclear testing can help resolve how the ocean mixes.

A complication acting in the opposite sense is the Suess effect, discussed also in chapter 4, where humans have diluted the atmosphere with respect to carbon-14 with carbon-12 from the burning of fossil fuels. The Suess effect also complicates the utility of carbon-14 in understanding ocean circulation; however, given the age of the deep ocean waters, the only signal will be in waters more recently exposed to surface exchange with the atmosphere.

Dissolved inorganic carbon in the deep sea, although isolated, still undergoes change. Bacteria and other microbes, protozoans mostly, metabolize any organic matter raining down from the surface layers as particles, or mixed down as dissolved organic substances, and thereby increase the quantities of dissolved inorganic carbon and dissolved inorganic carbon-14. The production of dissolved carbon-14 through respiration of organic matter raining down from above can significantly influence the measured age of the deep water.

■ ■ ■

Carbon dioxide, a gas, enters the ocean from the atmosphere. Unlike other gases, however, CO_2 undergoes a chemical reaction with seawater. The dissolved carbon dioxide reacts with the water to produce carbonic acid, or H_2CO_3. Carbonic acid then quickly splits—or, in chemical terms, dissociates—into hydrogen ions (H$^+$) and bicarbonate ions (HCO$_3^-$). Then the bicarbonate ions split into more hydrogen ions and carbonate (CO$_3^{-2}$) ions. (Like I keep saying, it's complicated.) CO$_2$, bicarbonate, and carbonate exist in a chemical equilibrium, depending on the water's temperature, salinity, and pH (how acidic or basic the water is). The dissolved carbon system also serves to buffer seawater against changes to its relative acidity. Normally, however, by far most of the carbon is bicarbonate. The ocean's carbonate system is large and makes the buffering system substantial, and also slows the exchange of atmospheric carbon with its ocean counterparts (the Revelle Factor).

Everywhere in the ocean there is a deficit of inorganic carbon in the surface ocean with respect to the atmosphere because of that slow exchange of carbon, and also because biological activity removes it. After the slow entry from the atmosphere, the CO$_2$ can stay as part of the reservoir of dissolved inorganic carbon in the ocean, or it can enter the ocean's food web. The first point of entry is photosynthesis, transforming CO$_2$ into organic carbon. Phytoplankton, the microscopic algae floating in the lighted surface layer of the ocean, carry out the bulk of ocean photosynthesis. Photosynthesis upsets the dissolved carbon equilibrium, but only briefly. The balance is quickly reestablished, albeit at a lower concentration.

The newly created organic carbon from photosynthesis continues its journey through the ocean's food web, from the phytoplankton to zooplankton to fish. The bigger particles of organic carbon can detour into jellies, protozoans, tunicates, and continue all the way to whales, at one end of the size spectrum, and to marine bacteria at the other. The fate of all these organisms is pretty much to get eaten, while slowly sinking downward. It's a cycle of eating, excreting, consuming again, excreting again, all in a downward trajectory. The ocean's food web is remarkably efficient in this regard. Only a small amount escapes to the deep ocean, beneath the thermocline, and an even smaller amount arrives at the seafloor. At every point, at each momentary reservoir in the food web, all protozoans, fish, bacteria, a copepod or krill, and diatom respire, putting inorganic carbon back into the water.

Carbon-14 makes up a slight but identifiable fraction of all these transformations and interactions. Theoretically, then, we can trace the carbon-14 atom in whatever happens to the dissolved inorganic carbon. The carbon-14 in the dissolved inorganic fraction might return to the atmosphere or be physically mixed down into the deep ocean. Likewise, carbon-14 can be traced into the food web and into the deep sea. Instead of being mixed downward by physical processes (although this can happen), excreta, the dead, and organic residues sink slowly to the abyss. Two endpoints to the ocean's food web are of interest in terms of carbon-14, one in an animal, the other chemical.

Many organisms in the ocean's surface layers make shells or skeletal structures, creating minerals, a process called biomineralization. Among the planktonic organisms, the minerals can be silica, in the case of the diatoms and Radiolaria, or even strontium sulfate for another protistan group, the Acantharia. Most commonly it is calcium carbonate, and the most common organisms making calcium carbonate shells are the coccolithophores and foraminiferans. Coccolithophores are tiny photosynthesizers that coat their cell walls with calcium carbonate plates called coccoliths. They produce massive blooms on occasion, making the water "chalky," which fisherman avoid, knowing that fish avoid the chalky water as well. The White Cliffs of Dover on England's south coast are almost entirely composed of ancient coccoliths that built up over millions of years and became consolidated into chalk. Before very fine meshed plankton nets were produced that could capture them, coccolithophores were thought to be an extinct group because the coccoliths, microfossils, were the only record of their existence.

Foraminiferans, single-celled protists related and similar to the familiar *Amoeba*, also make shells of calcium carbonate. Some make their living near the ocean's surface, along with the other plankton, and others crawl over the ocean bottom. Either way, foraminifera, nicknamed "forams," live for a few weeks or months and then die, the planktonic forms sinking and becoming part of the rain of particles to the deep sea. Both kinds, planktonic and benthic, end up in the sediments. I will discuss forams in chapter 13 for their importance to paleoclimate research with carbon-14.

The other endpoint for the ocean's food web is the dissolved organic residues resulting from all the various interactions in the plankton, and also from terrestrial sources via rivers emptying into the ocean.

Dissolved organic molecules make up most, by far, of the organic carbon in the ocean.

Unlike terrestrial ecosystems that produce mostly carbohydrates (think wood), the ocean's food web produces protein (think cod). My colleague Jennifer Cherrier uses the chicken soup analogy for the ocean's food web. Pieces of chicken, protein—the organisms—float in a thin broth of organics in solution. The organic residues dissolved in the "broth" are the molecules of life, but combined in complex ways. They are defined only operationally—that is, by mutual agreement. "Dissolved" is agreed to be anything that passes through a filter that would otherwise retain most of the smallest living things, the bacteria. That means the dissolved organics can be high-molecular weight molecules (including colloids), and small molecules, but also viruses. Dissolved organic carbon (DOC), being dissolved, will not sink. Instead currents and mixing, in and among the ocean's waters, transport and distribute DOC.

The ocean's dissolved organic carbon has the seemingly improbable combination of being both abundant and also unknown. DOC concentrations in the ocean exceed the carbon in living organisms by a factor of 200–300. Current estimates put the amount of DOC at 700 petagrams of carbon, while living carbon amounts to only 3 petagrams. (A petagram is 10^{15} grams.) At that amount, living organisms would appear insignificant, or even incidental, if they were not so active—growing, eating, excreting, and dying. By any measure, DOC is truly a sink, a repository, an endpoint for everything biological.

Like all carbon, the dissolved component can be isolated, concentrated, and fractionated by molecular size, and its carbon content measured. But what it actually is, its chemical composition, has, up to now, largely escaped determination. One clue is that the amount of DOC in the surface waters is double that in the deep sea, and because most biological activity occurs in these lighted depths, DOC must originate from marine organisms. Since the ocean's ecosystems—remember the chicken soup analogy—run on protein, and since DOC is such a sink, it stands to reason that those mostly protein pieces, as they decompose or are left to disintegrate, would be recycled, leaving the carbohydrate part behind. Indeed, much of the DOC seems to be sugars and other complex carbohydrates. The carbon-14 in the DOC, however, offers not only other clues but also mysteries.

First, the amount of carbon-14 in the DOC falls off rapidly going deeper in the ocean and arrives at a more or less constant value in the deep sea, below the topmost layers where most biological activity and exchange with the atmosphere occur. The carbon-14 in the DOC in the near-surface layers is continually replenished; the high values in the surface ocean suggest a biological source. Once "locked" in the deep sea, however, carbon-14 in DOC decays away the same way it does in the dissolved inorganic carbon and in the particulate organic matter. As expected, there are higher values of carbon-14 in the DOC in the deep Atlantic than in the deep Pacific, and the difference goes along with the age differences between the two oceans—about 1,000 years.

The problem comes when the ages of the DOC are determined. The carbon-14 ages of the DOC are old—really old: 4,000–6,000 years old. That means the deep-ocean DOC is old enough to have been traveling for several loops along the global ocean conveyor. The stuff does not seem to react or engage with anything else. It is an endpoint or a sink. Maybe some of the DOC upwells to a surface layer and gets hit by the sun's ultraviolet radiation, and might break down, but the ages of the DOC suggest that this is not an important pathway. Bacteria metabolize only a small but perhaps significant share of the old DOC.[11]

Another reason for the disconnect between the ages of the DOC and DIC is that DOC can come from rivers entering the ocean, or from seeps at the ocean bottom. Rivers carry DOC leached from trees and other higher plants. These are called "lignins" and "tannins." Tannins come, notably, from fir trees and make some lake waters dark brown, the color of tea.

Lignins and tannins are large, complex molecules, and probably hang around for a long, long time. Likewise, ocean bottom seeps can deliver really old carbon—for example, from shallow oil deposits. Hydrocarbons leaking into the ocean are three times as old as deep-sea DOC, at about 12,000 years old. Even though a minor component, these sources could skew the ages of DOC in the ocean's deeps. Getting samples of DOC at those ocean depths has not been easy; there have only been a few measurements at a few locations. So, right now, it is hard to figure out the impact of the deep-sea seeps on the DOC pool in the deep ocean.

Clearly, more data and more surveys could solve these problems. Remember, though, that the few data that exist demonstrate the difficulty of obtaining them. Ellen Druffel, at the University of California, Irvine, whom we met in chapter 7, has most of the data on the carbon-14 contents of DOC—a few profiles in each ocean, a few locations from which to explain the entirety of the largest store of carbon in the ocean.

11

OCEAN FERTILITY

The American Society of Limnology and Oceanography (ASLO), as it was known then, was convening its summer meeting at the University of Colorado in Boulder.[1] (Limnology, a not-too-familiar term, is the study of lakes.) I was standing in line for the offered food and drink at the usual Wednesday night social event, held at a nearby mountaintop ski resort.

Although lakes and the ocean are both watery parts of the planet, the "society" of limnology and oceanography was more like an arranged marriage. A small group of us were attending this particular meeting on the high plains and in the Rocky Mountains at the invitation of the president of ASLO at the time, Dick Barber, an oceanographer. Dick wanted to be sure there were at least some oceanography talks presented, so he put together a special session of speakers on ocean issues. Dick enlisted me as one of those speakers.

I am normally not one to initiate conversation, so that fell to the woman just ahead of me in the food and drink line. She asked me where I worked, and I said "the ocean." She said, "Wow, that must be really difficult to sample. So big." Taken aback, I murmured something like, "Yeah, it's tough." Then she asked what I worked on, and I replied "phytoplankton." She said, "Wow, those tiny guys must be really hard to work with." It had never occurred to me that my science punished me with obstacles both large and small. To me, oceanography was fun—serious work, but fun. Still, there was the unfortunate truth that I studied some of the smallest organisms, the phytoplankton, in the world's largest arena, the ocean.

The social event for me went downhill from there, helped along by the fact that I was the designated driver for the trip back down the mountain (another downhill) to the hotel in Boulder. I sat by myself to mull over this newly exposed reality, not burdened by drink, and not participating in any of the sociability of the event. I remembered a semi-drunken rant at a party from years before, by a more senior fellow graduate student, asking if I thought that being aboard a lone ship in the middle of the ocean and getting a single water sample of a few liters actually meant anything. (He used more colorful language.)

I realized that working on microscopic phytoplankton in a fluid environment that covers about 70 percent of the Earth's surface encapsulates the controversy surrounding the fertility of the ocean. Do we study the ocean itself, its dynamics and its contents, or do we focus on the behavior of microscopic organisms—in this case, the photosynthesizing phytoplankton? The tension between those who sample the ocean for physiological rates of the plankton and those who view the ocean from an ecosystem point of view, a more all-embracing frame, still exists today. The tension and controversy go back to the introduction of carbon-14 into oceanography in the early 1950s. Certainly there was oceanography—more specifically, biological oceanography—before the discovery of carbon-14, which is where I will begin.

■ ■ ■

In the 1930s and 1940s, formal education in oceanography was almost nonexistent. Oceanographers were all formerly ecologists, engineers, chemists, physicists, or in one case, a former budding theologian, and in another, an artist.[2] The ecology of the plankton—the plankton food web—was understood only in rough outline, limited, as always, by the available techniques used to measure ocean properties.

At the base of the food web are the phytoplankton: the primary producers, like those that Melvin Calvin and his Bio-Organic Group used as experimental subjects to figure out the cycle of photosynthetic carbon assimilation. (I reviewed the role of carbon-14 in unraveling the biochemical pathways of photosynthesis in chapter 6.) The phytoplankton convert inorganic carbon dioxide to organic matter using solar energy by photosynthesis, the main indicator of ocean fertility. Phytoplankton grow fast, doubling their numbers in a few days. If left unchecked, the otherwise clear blue ocean water would turn green or brown, as can happen in some unhealthy coastal and inland waters. Phytoplankton production, becoming phytoplankton biomass, must go someplace to keep the ocean blue, as we observe it.

That someplace is the zooplankton, the animal plankton that feed on the phytoplankton, and that in turn are eaten by small and larval fish, and those fish by larger fish (figure 11.1). Zooplankton encompass a wide variety of organism types, from microscopic, single-celled protists like *Amoeba* to

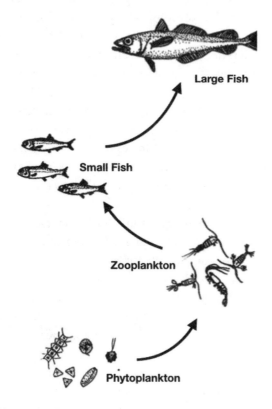

FIGURE 11.1

Simplified marine planktonic food chain as it was understood in the 1930s and 1940s, showing the phytoplankton (primary producers), a copepod (representing zooplankton), small fish, and a large fish. Phytoplankton are autotrophs, and zooplankton and fish are heterotrophs. (Bacteria, which break down the food web's waste products—the appreciation of which came much later—are not shown.)

copepods, krill, and shrimp. Fish that feed on zooplankton are small, such as menhaden, anchovy, and sardine, but zooplankton also feed Earth's largest animals. Filter-feeding whales and basking sharks subsist on krill, other small zooplankters, and small fish. Finally, there are the bacteria, which, as on the land, clean everything up and, in the process, break down organisms and their waste products to their basic molecular forms.

Sunlight to phytoplankton, phytoplankton to zooplankton, and zooplankton to fish sketches out a rough ecosystem organization, or what is sometimes called the trophic structure. "Trophic" refers to the manner in which, at each level, the organism or group gets its food. The phytoplankton are "autotrophs"—they feed themselves (aided by sunlight); the others are "heterotrophs," or "other" feeders. Applying some energy considerations, like how many calories are needed or how efficiently the trophic level can grow, you can see that if you know how large one part of the food web is, it is possible to figure out the others. For example, from the ocean's mass of baleen whales, including blue and finback whales, we can put some numerical bounds on the primary, or photosynthetic, production needed to support that biomass. From the other end, the potential for fish production in the sea could be (and has been) estimated from the penetration of sunlight into the ocean and the photosynthesis by phytoplankton.[3]

In those early days, during the 1930s and 1940s, understanding how the ocean worked ecologically meant taking measurements of the biomass of the food-web components: the phytoplankton, the zooplankton, and the fish. Rarely, photosynthesis was measured directly using changes in oxygen. (Recall from chapter 6 that photosynthesis gives off oxygen through the energy gained from splitting the water molecule, and uses that energy to assimilate carbon dioxide.) The analytical techniques for measuring oxygen in water were developed in the nineteenth century by Ludwig Wilhelm Winkler in Budapest. He published his method in his doctoral dissertation, using it to measure the oxygen content in lakes. "Winkler bottles" are still available to measure biochemical oxygen demand (BOD) and are used in water-quality studies.[4]

Winkler's method worked well in inland waters and estuaries, but not in the ocean. It did not have the sensitivity to measure the changes in oxygen from phytoplankton photosynthesis, except under unusual circumstances. The changes in the ocean were too small. Oxygen changes also resulted from respiration from the zooplankton and bacteria, which could not be separated from that of the phytoplankton. The direct measurement of phytoplankton production was relegated to a minor role in the more important measurements of biomass of the trophic components.

Gordon Riley (figure 11.2), working at the Woods Hole Oceanographic Institution (WHOI), formulated some of the first collective relationships among the plankton. While his oxygen flux experiments for photosynthetic production figured into the analysis, he also considered how much primary productivity would be needed to support the zooplankton biomass that he measured. The direct measurements of photosynthesis were a way of validating his results from trophic analyses.[5]

It was pretty clear that Riley would never be able to get enough measurements of the systems he studied; the sampling problem that my grad

FIGURE 11.2

Gordon Arthur Riley.

From the Manuscripts and Archives, Yale University Library, in Frank N. Egerton, "History of Ecological Sciences, Part 58B: Marine Ecology, Mid-1920s to About 1990: Carson, Riley, Cousteau, and Clark," *Bulletin of the Ecological Society of America* 98, no. 2 (March 2017): 113–149.

student colleague ranted about was ever-present. So to "connect the dots" (his data points), he created models of the trophic relationships and how those relationships played out over the seasons in the temperate ocean and coastal waters. Riley himself admitted that his models were long on ideas and short on supporting data. Nevertheless, they have endured. Overall, they reflected a view of the ocean's ecosystem as dynamically interacting, and needed to be evaluated as such. Thus, his efforts were extensive rather than intensive. They dealt with the ocean ecosystems—interacting populations—and not with processes by the trophic components in isolation. The introduction of carbon-14 to oceanography in 1952 changed all that, to focus on one trophic element and process: phytoplankton photosynthesis.

■ ■ ■

The introduction of carbon-14 to oceanography came not long after carbon-14 was being used to work out the assimilation of carbon dioxide in photosynthesis (chapter 6). In 1951–1952, Einer Steemann Nielsen (figure 11.3), a Danish marine botanist, described a method for adding carbon-14, as isotopically labeled sodium bicarbonate, to seawater samples. The idea was that carbon-14 would be taken up in the seawater's particulate matter, of which the phytoplankton were a part. Then, after an appropriate length of time, the particulate matter from the seawater would be separated out and assayed for radioactivity in a Geiger-Müller counter. The method would be a marriage of the pathway of carbon in photosynthesis worked out at Berkeley with methods used for carbon-14 dating.

Adding the carbon-14 as bicarbonate might seem unusual from what we know of the chemistry of photosynthesis (again, chapter 6). However, the inorganic carbon species in seawater—dissolved carbon dioxide, bicarbonate ion, and carbonate ion—exist in a balance, a chemical equilibrium. Adding a small amount of carbon-14-labeled bicarbonate does not disturb that equilibrium, and arriving at the equilibrium means that some of the carbon-14 bicarbonate will become carbon-14 in the dissolved carbon dioxide. If the phytoplankton remove carbon dioxide through photosynthesis, the chemical equilibrium among the ions and dissolved carbon dioxide is quickly reestablished. In any event, as I pointed out

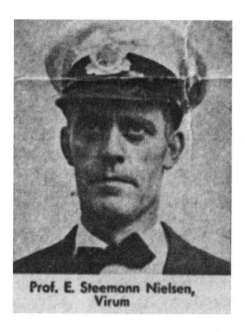

Prof. E. Steemann Nielsen, Virum

FIGURE 11.3

Einer Steemann Nielsen.

in chapter 6, storing isotope as bicarbonate, as a powder or dissolved in seawater, is much easier than storing and using carbon dioxide, a gas.

The application of carbon-14 to ocean photosynthesis developed in a certain amount of obscurity. At least, Steemann Nielsen never directly described the source of his idea. During the 1940s, he was Professor of Botany at the Royal Danish School of Pharmacy. He was aware, from others, of the initial work on carbon-14 at both the University of Chicago and the Berkeley Radiation Laboratory. Three people converged in Copenhagen in the late 1940s: Lise Schou, Hilde Levi, and George von Hevesy.

Lise Schou worked as a student with Calvin's group at Berkeley. She was first author on one publication and coauthored another in the group's series of publications on their progress in working out the path of carbon in photosynthesis using carbon-14. Schou returned to Copenhagen in 1950 with experience in the use of carbon-14 in photosynthesis studies and presented a seminar on what became her doctoral dissertation at UC Berkeley.

Hilde Levi, from Berlin, had a doctorate in physics and was a colleague of George von Hevesy in Copenhagen. Von Hevesy received a Nobel in Chemistry in 1943 for his work on biological tracers. Levi had previously spent a year in Libby's lab at the University of Chicago, and became acquainted with low-level counting of isotopes such as carbon-14. Levi returned to Copenhagen in 1948. (Later, she contributed the technical aspects to the method as used at sea.) Thus, as noted by Dick Barber and Anna Hilting (2002; see also Sondergaard 2002), Steemann Nielsen had knowledge of the value of biological tracers, the use of carbon-14 in photosynthesis of microalgae, and knowledge of how to count low levels of isotope. The remaining step was to join these three pieces of knowledge and apply them to oceanography.

Steemann Nielsen took his new method on the road—or, we should say, to sea—in 1950, on a worldwide expedition sponsored by the Danish government, on board a ship called the *Galathea*. The trip became known as the *Galathea* Expedition, one of a series of worldwide surveys sponsored by European countries, including the first *Galathea* Expedition in 1845–1850, the *Challenger* Expedition in the 1870s, and the *Meteor* and *Dana* cruises in the 1920s.[6]

The "organic productivity" of the sea was not a central focus for the *Galathea*; the Expedition was more a "cruise of opportunity" that allowed Steemann Nielsen to test his new method for measuring productivity using carbon-14. As with other nationally sponsored expeditions, the plan was for the *Galathea* to stop at various locations for oceanographic observations, and then also to conduct measurements of photosynthetic carbon assimilation using carbon-14.

Any worldwide seagoing expedition might take months or years to complete, with "worldwide" in reality meaning a visit to the various oceans. All such expeditions confront the sampling obstacle mentioned above. The sheer immensity of the ocean means, in effect, astonishingly few locations where Steemann Nielsen's measurements could be made. The *Galathea* sampled more intensively near coastlines. It stopped only a handful to a dozen times in each of the open expanses of the Atlantic, Indian, and Pacific oceans, neglecting, by logistical necessity, vast ocean areas (figure 11.4).

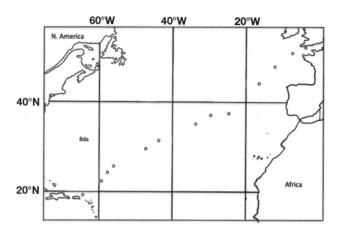

FIGURE 11.4

Open ocean stations (indicated by "o") occupied by the *Galathea* Expedition (1952) in the North Atlantic Ocean. Station positions are not exact (see Steemann Nielsen and Jensen 1957, fig. 19). The approximate location of Bermuda is denoted by "Bda."

At each of the *Galathea*'s locations, the researchers used Steemann Nielsen's "carbon-14 method," as it came to be known.[7] The method is simplicity itself. Sterilized glass containers, called ampoules, are filled with a solution of sodium bicarbonate, with a small amount of the bicarbonate having carbon-14 as its carbon atom. In chemical jargon, this is $NaH^{14}CO_3$. Before a cruise, the ampoules are filled with isotope solution and sealed by melting the glass at its top end. Then they are sterilized and stored until used. For the experiment, the ampoules are broken and their contents added to a seawater sample in a clear glass bottle. The seawater contains natural, or wild, plankton populations from all the trophic groups: phytoplankton, microscopic zooplankton, and bacteria, and sometimes a larger member of the plankton, a copepod.

The sample, in a clear bottle, is illuminated, and both inorganic carbon-12 and carbon-14 are taken up by the phytoplankton through photosynthesis. After a length of time, the contents of the bottle are passed through a filter, with the filter retaining the particles, essentially all the plankton. Usually, the filters are small disks of cellulose or glass fiber.

The disks also become wet with the seawater, and any isotope not taken up into organic matter will be in the wet filter as well. To eliminate the isotope that was not incorporated into the particulate matter, the filter is exposed to acid fumes. The acid environment is strong enough to convert any bicarbonate to carbon dioxide, a gas that is given off, but not so strong as to break down the organic matter in the particles. The filter can then be assayed for its radioactivity either in a Geiger-Müller counter or, more recently, in a liquid scintillation counter. The amount of radioactivity, the carbon-14 assayed, is proportional to the photosynthesis occurring in the sample. If you know the amount of isotope appearing in the particulate matter in proportion to how much isotope was added, and also the amount of bicarbonate already present in the seawater, the rate of photosynthesis is easily calculated.[8]

If these essentials of the carbon-14 method do not sound simple, they really are. Not only is it simple, carbon-14 radioactivity can be assayed with extreme sensitivity, as shown in chapter 7. And carbon-14 is the perfect isotope. At the levels used, it is safe, requiring no special apparatus or safety features. Whether being prepared as bicarbonate or carbon dioxide, it is closer to a waste product for humans than anything metabolically important, so using it for assessing ocean photosynthesis will not incur any harm. It is used overwhelmingly by photosynthesizers in the ocean and not by other organisms in the plankton. It can be added in trace amounts, and because there is so much bicarbonate already in seawater, adding carbon-14 as bicarbonate does not change the chemistry or interfere with other biologically important components or interactions. For measuring photosynthesis in the ocean, carbon-14 is an ideal tracer.

In a certain way, then, it is surprising that the "carbon-14 method" attracted such controversy. There were worries about contamination. How chemically "clean" is the bicarbonate solution to be added to the seawater sample? How "clean" are the bottles that contain the plankton over the period at which they are incubated experimentally? Ocean waters—at least in the recent past—are some of the most pristine found on Earth, and can therefore be easily altered or adulterated. How free from contamination are our water collection devices? Do they change the water samples chemically? Is it even a good idea to trap and enclose a quantity of seawater, removing the organisms from their natural habitat? Is the trophic structure of the ocean that is being sampled adequately represented during the experiment, or does containment disrupt it? These issues sometimes mentally exhausted those attempting the assessment.

Yet the ease with which photosynthesis measurements could be made at sea with Steemann Nielsen's method won the day. The carbon-14 method attracted teams of oceanographers eager to find out how ocean fertility, marked by the rate of photosynthesis, varied geographically; varied with other oceanographic phenomena; and changed with the seasons or from day to day, or even within a day.

The simplicity of the method hides significant problems that go beyond the contamination or containment (incubation) issues. First, because the method is so easy, a large number of measurements can be made. Second, the method has unmatched sensitivity for even very low rates of photosynthesis in the ocean's "deserts." Third, the method measures appearance of carbon-14 in organic matter. Only a positive outcome can occur; there is no such thing as a negative incorporation of carbon-14. Together, these attributes mean that one can collect lots of data with lots of measurements, be consoled by universally positive results, and have those results stand by themselves. The extreme sensitivity of the method and the small rates that exist in most parts of the ocean mean that results from the carbon-14 method cannot be compared or validated against other kinds of estimates of fertility or productivity. Herein lies a trifecta of a problem in understanding ocean fertility.

The introduction of carbon-14 into oceanography happened at a critical moment. So confident was Steemann Nielsen in his assessment of ocean fertility that he embarked on a small crusade to disparage the work of others, particularly on the North American side of the Atlantic. Before the introduction of the carbon-14 method for measuring aquatic photosynthesis, the only other means, as I have pointed out, was to measure changes in oxygen, a by-product of photosynthesis.

Winkler's method was marginal at best in the ocean. It takes a lot of photosynthesis to nudge the oxygen level enough for the Winkler technique to see it. The remedy was to keep the experiment, the incubation of water, going for long periods so that oxygen would change. Of course, at the same time, the planktonic organisms—phytoplankton, protozoans, bacteria—respire away, consuming oxygen at the same time it is being produced. So, the "net" amount of photosynthesis can be close to zero, or on many occasions over a day, zero.

Needless to say, oceanographers in those days would not rely solely on the oxygen data, but would use other kinds of information to understand how productive a region of the ocean might be and how that productivity is regulated. Since his was a direct measure of photosynthetic carbon assimilation, I am guessing that Steemann Nielsen probably felt he did not need much information regarding how nutrients are supplied, the effect of zooplankton, temperature, and other environmental constraints. His method went to the heart of the matter, straight to the photosynthetic assimilation of carbon.

When the *Galathea* Expedition measured the photosynthesis in the open ocean, and specifically in the Sargasso Sea south of Bermuda (see figure 11.4), Steemann Nielsen found very low rates of photosynthesis. A few years before Steemann Nielsen introduced the carbon-14 method, Riley, together with Henry Stommel, a physical oceanographer, and Dean Bumpus, a chemist, published a grand synthesis of the biological oceanography in the Northwest Atlantic, and perhaps tellingly, north of Bermuda. The work still stands today as a hallmark, and as one of the founding publications of biological oceanography (Riley, Stommel, and Bumpus 1949). To estimate photosynthesis, Riley relied on oxygen fluxes, as he had many times in previous years. Combining those with the physics and chemistry of the system, and an understanding of the relationships among phytoplankton and zooplankton, the team was able to formulate ideas underlying the production ecology of the planktonic system.

The rates of photosynthesis Riley measured were at times 10-fold higher than Steemann Nielsen measured with carbon-14. Earlier, Riley (1944) had published a short paper with the grandiose title "The Carbon Metabolism and Photosynthetic Efficiency of the Earth as a Whole," in which he estimated global ocean productivity at 126 gigatons per year.[9] Steemann Nielsen, based on the transit of the *Galathea* through the global ocean, estimated it at 20–30 gigatons, about an order of magnitude less.[10] So, which is it?

Steemann Nielsen attacked Riley, saying his numbers must be wrong, and implying that some other process was being measured in his incubation bottles, perhaps a "bactericidal" effect. Steemann Nielsen continued his diatribes in several publications. But Riley had a couple of allies at WHOI in John Ryther and Ralph Vaccaro, who published work directly countering Steemann Nielsen's criticisms.[11] Riley defended his analyses in 1953, in a letter he wrote to the editor of the same journal where Steemann Nielsen had published his attack (Riley 1953). The controversy simmered at a low boil over those years, without a definite resolution either way.

Finally, Riley, in 1957, reported data collected weekly for two years from a weather ship stationed in the North Sargasso Sea. Weather ships—actual weather stations at sea—have a prominent history in oceanography, supplying data collected at a geographic position over time, a "time series," for locations that would be impossible to sample otherwise. (That tradition lives today in two ongoing time-series stations near Bermuda and Hawaii, sponsored by the National Science Foundation.[12])

Riley used three independent lines of evidence for his assessment of productivity: seasonal changes in oxygen in the water, over depth; the demand for productivity by the zooplankton; and oxygen flux measurements in contained water samples. Within the ability of the methods available at the time, the three more or less independent methods gave consistent results and agreed with previous values. Riley's results from his two-year study exceeded by a factor of three Steemann Nielsen's values from the three locations the *Galathea* Expedition occupied in the central North Atlantic.

Biological oceanography went through an unhappy time in the 1950s. The back-and-forth between Riley and Steemann Nielsen never reached the same level of bitterness as the previous fight about ocean plankton between the German biologists Victor Hansen and Earnest Haeckel in the late nineteenth century.[13] But it was still not a happy time, and foreshadowed decades more of controversy, uncertainty, and misunderstanding. However, the sensitivity and ease of the carbon-14 method proved too great, and Steemann Nielsen prevailed.[14] The 1960s saw many, many observations using carbon-14 in lakes, estuaries, and the ocean, and even in large plastic spheres submerged below the water's surface. One of those spheres, holding about 3,000 gallons, was held next to a pier and filled with water filtered free of particulate matter. A small volume of natural seawater was added, along with carbon-14, and the incorporation of the isotope into the phytoplankton was followed for two weeks.

The carbon-14 measured about the same change as the increase in phytoplankton cell carbon.[15]

By the 1970s, oceanographers began drawing maps showing the variability in primary production, as estimated by carbon-14, across the oceans. But the concerns about carbon-14 in biological oceanography nagged and persisted. The conflicts came from many quarters and mostly had to do with squaring carbon-14 primary productivity results with activities of the other parts of the planktonic realm, reminiscent of the battles between Riley and Steemann Nielsen in the 1950s.

Bacteria, like everything else in the ocean, rely on the organic matter produced by primary production—and the bacteria presented the first conflict. Their use (respiration) of organic carbon residues and wastes from the food web seemed to exceed the amount that could be supplied if the carbon-14 values for primary production were correct. Carbon-14-based primary production could not match the respiration measured in the bacteria. One big related problem was that early on bacterial cells in the ocean were severely undercounted.

Up until the late 1970s, bacteria were estimated by the number of colonies that grew up from small water samples placed on bacterial growth media in petri dishes, each colony representing one bacterial cell from the ocean. These so-called "plate counts" showed bacterial numbers to be low enough to be not that important in the overall scheme of things—a big mistake. Improved microscopic techniques in the late 1970s showed millions more bacteria than earlier reported.[16] The mistake was to assume that all bacteria found in the ocean will grow on prepared laboratory media. Only a very small fraction of the bacteria types found in the ocean will grow in a petri dish. Their numbers were badly underestimated. The much higher numbers and activity of marine bacterial populations identified in the 1970s had to be supported by primary production. Oceanographers were now finding it difficult to summon enough of it to support the newly recognized bacterial numbers.

The products of photosynthetic production eventually sink to deeper depths, and oceanographers can catch that material in devices similar to rain gauges, called particle interceptor traps. Many oceanographers refer to the flux of the sinking matter as "the rain rate," where the "raindrops" refer to particles of organic matter and not precipitation, sort of the reverse of what we experience on land. The amount of material caught in the traps seemed to be more than could be explained, again, by the amount of plankton produced near the surface, using the values given by the carbon-14 method.

Other researchers looked at the increase in oxygen over the summer months as an indicator of oxygen-evolving photosynthetic production, and again, the carbon-14 numbers compared much lower. Still others estimated the utilization of that same oxygen by microbes, respiration, in the deep sea and concluded, yet again, that ocean photosynthetic production must be much higher to support that. Those studying fish populations entered the fray, saying there was not enough photosynthetic production to support the number of fish being caught. Even the issue of contamination of the seawater samples resurfaced. Did we collect them cleanly? Were our isotope additions clean? Were the bottles we used for containing the plankton populations free of contaminating substances? And on, and on, and on.

In many of these studies, if not all, the various investigators worked in isolation. The bacteriologists went to sea on their own cruises, as did those who studied the phytoplankton and made the carbon-14 photosynthesis measurements. The researchers doing measurements on the water itself did their work separate from others and could only compare their results to carbon-14 data from other cruises, locations, seasons, or years.

That brings us to the early 1980s, when the various conflicts surfaced and became topics at workshops and conferences. The carbon-14 method had become everyone's favorite "whipping boy." All of these studies to some extent reprised the controversy begun by Steemann Nielsen in advocating his carbon-14 method. Riley, with his trophic analysis of biomass changes, was on the other side, more akin to those raising issues based on oxygen changes in the ocean water itself, the demand for primary production by bacteria and fish populations, and the rain of particles falling out of the surface layer of the ocean.

Dick Eppley, from the Scripps Institution of Oceanography, thought about those disparities in how the ocean's biology worked and conceived a project called "Plankton Rate Processes in Oligotrophic OceanS," or PRPOOS (and pronounced "purpose"). He became the project's principal investigator (PI) and chief scientist for the seagoing part. In its early days, doing the research in Hawaii, we gave him the name "Magnum-PI," after a television show based in Hawaii that was popular at the time.

The seagoing part, the PRPOOS cruise in 1985, represented a culmination of the work in ocean fertility that had gone on for the previous 30-plus years. The PRPOOS cruise brought a record amount of carbon-14 to sea for 30 days of experimentation to put to rest the controversy surrounding the method, and ultimately the question of photosynthetic production in the ocean. Although it was too much to expect that the PRPOOS project would solve the problem of ocean fertility on that one cruise—30 days for a 30-year controversy—those of us who participated made substantial advances, still recognized in biological oceanographic research. Our results, if representative, probably doubled the previously assessed ocean's biological productivity. How that happened is the topic of the next chapter.

12

RESOLUTION

———

Plankton Rate Processes in Oligotrophic Oceans

"**W**e're done for," I thought.

The RV *Melville*, an oceanographic research vessel operated by the Scripps Institution of Oceanography, was just tying up to the pier in Honolulu, Hawaii, after 30-odd days in the North Pacific Central Gyre. Waiting on the pier was a woman—obviously not a local—looking all business. She had a determined look and held tightly to the handle of a stainless steel briefcase.

Our arrival in Honolulu had been the culmination of two years of planning and almost endless wrangling over radioisotopes, carbon-14 in particular. We use minute quantities, microcuries, of the isotope in each of our experiments. Still, these amounts are millions of times higher than what exists naturally in the environment. For this cruise, we had brought a world-record amount of isotope to sea, more than any previous expedition, creating jitters, worry, and anxiety throughout the oceanographic community. All told, we had about 100 millicuries, maybe a 1,000 times more than was typically carried aboard ship. This was unheard of.

To say that our 100 millicuries caused consternation with Ship Operations at Scripps was an understatement. Others who use research vessels for sampling the natural abundance of carbon-14 in the ocean, such as on a GEOSECS-type cruise (see chapter 10), dreaded the thought of the *Melville*'s becoming contaminated forever (well, for 40,000 years or so). Still, the National Science Foundation (NSF) had funded our project and the ship time. Our proposal was fully reviewed by our scientific peers, followed by further review by a panel of experts, all of whom recommended

funding, and we were finally given the go-ahead by the NSF's Ocean Sciences Program.

During the cruise planning, we established special protocols, many
used for the first time at sea. All isotope work was confined to portable
labs, essentially cargo containers, placed on the weather decks and outside the superstructure of the *Melville*. We took extreme care in handling
the isotopes—not only carbon-14, but also tritium (hydrogen-3). Dick
Eppley (figure 12.1), the chief scientist, and his helpers instituted what
we first thought were horrendous and overbearing procedures that might

FIGURE 12.1

R. W. (Dick) Eppley on the PRPOOS cruise, 1985.

Photo by J. F. Marra.

stifle our science. Each of the on-deck portable labs had an anteroom, used to don booties and lab coats. We wore surgical-style latex gloves. All countertops were covered in absorbent sheeting with plastic backing to contain any water. We used lots of paper wipes; after use, those wipes and anything else that even possibly came into contact with radioactivity were put in special trash containers for later disposal.

Every five days while we were at sea, we isotope users followed Dick around the ship, swabbing doorknobs and handles, the deck areas where we thought the "contaminators" would step, and the countertops, putting these little samples in vials to be assayed later. The "swabs" were inch-diameter discs made of glass fiber, which we wetted with distilled water and, using tweezers, rubbed on the various surfaces. It was burdensome to some of us, and maybe even slightly ridiculous. We grumpily complied— it was part of the deal.

We did not realize how puny our efforts were until the official-looking woman from the pier came on board, even before U.S. customs inspectors. I watched the woman enter one of the isotope labs on deck and then return outside almost immediately. She opened her case on deck, donned plastic gloves, and pulled out a large squeeze bottle and what looked to be a small yellow disk. She squirted a liquid from the squeeze bottle over about a three-foot square of deck surface just outside the lab door, and it foamed up. I guessed the liquid was a kind of detergent. Once the deck was soaked, she threw down the small yellow disk, and it immediately inflated to a large sponge. I was reminded of those childhood toys, tiny blots that, when you put them in water, swell into various sea-animal shapes. This was a big animal. The sponge absorbed all the liquid on the deck; the woman picked it up, and squeezed the liquid from the sponge into another large, wide-mouthed bottle. And we thought we would find some radioactivity with our little glass fiber discs?[1]

We found out later that the expansive and expensive assays to check for contamination of the ship had turned up nothing. With our record amounts of isotope, all our isotope handling, and all our experiments, we had not contaminated the RV *Melville*.

The *Melville* was the seagoing part of the project I mentioned in chapter 11 called Plankton Rate Processes in Oligotrophic OceanS, or PRPOOS (pronounced "purpose"). PRPOOS encompassed years of planning and preparation, building up to this just-completed 30-day cruise in late summer 1985. We call these seagoing trips "cruises." Invariably, when I would tell my nonscientist friends that I was going on one, they would say "Oh, how nice, a *cruise*," thinking Royal Caribbean. Not quite.

Work on oceanographic "cruises" is around the clock, every day—no weekends, no real time off. You want to squeeze in every scrap of time for work. The PRPOOS cruise in 1985 was not only a finale, but also a culmination of research into the use (and misuse) of carbon-14 in oceanography that had gone on for more than 30 years. PRPOOS was the first open-ocean investigation in which we could compare various ways of measuring fertility—specifically, the primary, or photosynthetic, productivity of the ocean. As I related in chapter 11, in those previous 30 years, carbon-14 had played a starring, maybe an infamous, role.

Four events on the PRPOOS cruise had the potential for upsetting things and, in a couple of cases, for utter calamity. But each time, the incidents finished well; disaster was averted, and one case led to a positive outcome.

At some point during the planning of the cruise, Chief Scientist Dick Eppley made an announcement to us, his small planning committee—"the three Johns": John Sieburth, John Heinbokel, and me. Dick said that he was arranging to bring his wife, Jean, along on the cruise. We were skeptical, if not incredulous. Bringing along a nonscientist spouse on a cruise in those days was frowned upon. Berths aboard oceanographic vessels were at a premium. Everyone wants an extra person, but space is limited, as are opportunities for work at sea.

Then we thought about it. We all knew how close Dick and Jean were. And Dick at sea without his wife might be, socially speaking, perilous to shipboard harmony. He could get cranky. We eased off our concern, especially when he showed us the redesigned chief scientist's cabin, with a newly installed double bed. Jean would not be taking up anyone's extra berth. And she proved to be just what the cruise needed. She became a great ambassador from the scientists to the ship's captain and his mates and, even more importantly, to the cook, who, with Jean's encouragement, readily supplied more cookies and snacks between meals. I remember sitting next to her during one lunch and asking, "Well, Jean, this is your first long cruise after all these years. How do you like it?" She immediately replied, "I haven't had to cook one meal!" She was also a first-rate tech, helping to haul and filter seawater for important analyses, and generally making herself available to assist. It ended up being a great decision to bring her along.

Usually, aboard ship, a "man overboard" event is a drill, a practice for the very rare real thing. What happened on the PRPOOS cruise *was* the real thing.

Most of the work on a cruise where biological measurements are being made occurs at night. Yes, we need sunlight to illuminate our experimental subjects, so they can photosynthesize. However, getting the samples to that point means getting up at 2 AM to begin to collect all the water and prepare the day's experiments. By dawn, most of us have only enough energy to get the morning meal from the galley before catching a few hours' sleep. During the day, the ship's decks are quiet; the only sound, aside from engine noise, is the burbling of the water pumped through our on-deck, seawater-cooled tanks used to incubate samples. The daylight hours also inflict boredom on the crew up on the bridge driving the ship, especially on this cruise when we would sit at one place for days at a time.

I cannot remember why, but I happened to be on deck that day, checking on things. Around noon, the bridge watch decided to move the ship, not with an easy acceleration, but with a real foot on the gas, kicking the ship forward and, for a bit of extra panache, adding an abrupt turn to port. It just so happened that the ship's marine tech was adjusting a sunlight sensor atop one of the portable deck vans near the stern, on the starboard side. At the quick acceleration and port turn, he lost his balance and fell into the sea.

Elijah Swift, an expert on marine bioluminescence, who normally worked at night, had for some reason decided to take a break from his analyses for a stroll on an upper deck of the ship, just above the bridge. His eye caught the marine tech's plunge into the water. Elijah shouted, "Man overboard!" At that point, I was checking on the water tanks on the rear deck. When I heard Elijah, I went to the starboard side, looked toward the horizon, and saw a red dot a couple of miles from the ship. I say "red dot" because the marine tech's head was both shaved and sunburned, and his head was all that was visible above the waves. It stood out against all that blue. I grabbed a life preserver and ran to the bow, ready to toss it overboard. By this time, the ship's captain was on the bridge, maneuvering the *Melville* to pick the marine tech out of the water. The captain shouted to me *not* to toss the life preserver, so I brought it back to its rack, my attempt at heroism deflated.

An engineering innovation for the class of oceanographic research vessels that included the *Melville* was a unique propulsion and steering system. Instead of the typical stern "screw," the *Melville* had "cycloids," a combination of thrust and steering, affectionately (or not) called "egg beaters." The cycloids consisted of a circular array of small rudders. Changing the pitch of the rudders moved the ship forward and back, and even side to side. How fast the cycloid array was spun around gave the ship its speed. *Melville* was not a speedster; the cycloids were not designed for speed. But they made the ship incredibly and gently maneuverable. The captain eased the *Melville* up sideways to the tech, who was treading water. The crew put a sea ladder over the side, and he clambered up. Back on deck, he was philosophical, and even joked about his time in the drink. We all went back to work, maybe a little more chastened by what could have happened if Elijah had not taken that break for fresh air.

The third event made me happy, even ecstatic, to have done my isotope work in a deck van, an "isolation" lab.

As I said at the beginning of this chapter, the portable deck labs had anterooms for donning booties and lab coats before entering the work area. One problem, however, was that our experiments were done outside the lab, mostly in small Pyrex or Teflon bottles. Sometimes we made microscope slides from samples that had been spiked with carbon-14. One other kind of experiment, however, involved placing a huge plastic bag in the water alongside the ship, sort of like the plastic sphere experiment in the early days of using carbon-14 that I talked about in chapter 11.

Years later than that plastic sphere, and some years before PRPOOS, Dutch researchers had used large plastic bags towed behind a ship and had come up with unexpectedly high rates of photosynthesis—a result that had to be checked. Maybe our normal incubation bottles were too small to be representative of everything happening among the ocean's plankton? The idea, then, was to test the "sample size" problem by using, similarly, a large, transparent plastic bag with a volume of a few hundred liters. Curt Burney, from Nova University in Florida, devised a system whereby the bag was put in the water and, with a strong jerk on the line, sprung open and filled with surface seawater. With another pull on the line, it closed, capturing a few hundred liters of seawater. One way to think about this is a baleen whale scooping up a volume of water, except without the baleen to filter out the krill. The closure was fitted with a length of tubing that allowed the bag to be sampled.

We did one or two test runs. The actual deployment could not be done from the deck of the ship; we were too high above the water. The bag and its tubing had to be prepared entirely in a small boat, tethered to the side of the ship and subject to waves and heaving seas—not fun. Setting up and deploying the bags was a major effort on Curt's part, but he was the only one who knew how to do it. We could only stand by helpless, as the small boat bobbed up and down while Curt made all the connections and muscled the bag over the side. By the end, he was clearly exhausted.

Curt is a member of that class of oceanographers who like to keep things simple and basic. Their slogan is "No electrons, no moving parts, works every time." There's a lot to commend in this point of view. Curt at least kept to the "no electrons" part of the phrase, in what happened. For this particular experiment, we were going to spike the bag's contents with carbon-14. The bag would be, for us, a very large "bottle" to test whether the size of our experimental samples affected the photosynthetic rates we measured.

After he deployed his big bag from the side of the small boat tied to the *Melville*, Curt ran the plastic tubing up onto the deck, tying it along the railing and passing it through a port and into our van. Inside the van, the plan was to pump the carbon-14 spike into the bag via the plastic tubing; we were also to pump samples out of the bag, at selected intervals, using that same tube. The experiment was to record the time course of the appearance of carbon-14 in the particulate matter, giving us the rate of photosynthesis.

Conforming to the "no electrons" idea, the pump action and driver turned out to be Curt's arm. He rigged up a hand-cranked pump that worked by peristalsis, the mechanical parts of the pump sequentially squeezing the tubing to move the water through it. The process reminded me of those old silent movies where when the Model-T breaks down, someone gets out, goes to the front of the car's radiator, and uses a hand crank to turn over the engine's pistons to get it going again.

Curt threaded the tubing through the pump's workings and put a plastic connector on the end. To that connector, he attached another length of tubing, and this went into a plastic gallon jug. Curt started cranking the pump, and soon we had water from the bag going into the jug. When we had enough, Kristina, my tech, prepared the carbon-14 spike and added it to the jug. So far, so good. Then we put the tube back into the gallon jug, and Curt reversed his cranking direction, sucking the seawater out of the jug, and sending it through the tubing, out of the van, and back into the bag over the side of the ship. We used another jug-full of seawater pumped into the bag to flush the tubing. We then had to wait until we thought the spiked sample we had pumped back into the bag was mixed with the rest of the unspiked seawater. If you think about it, this does not take long; the bag is continually moving, being jostled about by the ocean waves and currents and the lines used to secure the bag to the small boat.

Perhaps it was Curt's exuberance. He reversed the cranking to take the first sample—the first time point. But all of a sudden the tubing came unfastened, and like a loose fire hose, it wildly sprayed the inside of the van and all of us standing by, waiting expectantly. We subdued the hose end, but not before we had been splattered with radioactive bag water. To be honest, Kristina and I differed on the perceived dangers of carbon-14. To say she was agitated would be a gross understatement. I can't say as I blamed her.

In a few seconds, we were able to pinch the offending tubing, stopping the flow, but that ended things for a while. Fortunately, the carbon-14 method is extremely sensitive. We did not need to add much to conduct the experiment, so the water that came out of the bag was not that "hot." Still, we were all upset, but got down to the task of cleaning up the inside of the van to remove all traces of our "crime"—disposing of our lab coats and booties and donning new ones, and swabbing to ensure all the surfaces were at background radiation levels. We also continued with the experiment, recording the carbon-14 assimilation over the next couple of days, and successfully recovering the bag.

The farcical tubing break was not the only disconnect involving the spiked bag. Dick Eppley was concerned that there might be some carbon-14 activity in the bottom of the small boat used to launch and recover the bag, and where the tubing was attached. So he dispatched his tech, Ed Renger, to investigate. To sample the water in the bottom of the boat, Ed took with him an automatic pipettor. Automatic pipetting devices have now become ubiquitous in all kinds of labs. In those days, they were still relatively new, and oceanography, without the massive needs for consumables of, for example, biomedical research, always seemed behind in lab technology. Anyway, instead of using your mouth to suck up reagent, solvent, or some other noxious substance, as with an old glass pipette, the pipettor only requires a press of the thumb to expel the air in a small plastic tip, and then releasing the thumb after the tip is inserted into the liquid to suck up a prescribed, usually very small, volume. For repetitive sampling, taking small measured amounts, or spiking a series of containers with isotope to be used in an experiment, pipettors are ideal.

Ed jumped into the boat with the auto-pipette, supplied by Don Redalje, and, with gloved hands, took the water sample and immediately delivered it (with another press of his thumb) into a scintillation vial. Ed climbed back aboard, and Dick took the sample for liquid scintillation counting.

The sample showed significant radioactivity! Was the water in the boat, not to mention the boat itself, contaminated with carbon-14? Ed, having been in the boat, was immediately hosed off at the ship's rail, and the deck area flushed with seawater. We were all stunned and perplexed. And we also needed more samples for verification.

After taking more samples, much hand-wringing, and looking at the problem from all angles, we finally discovered it was not the boat that was contaminated. It was the pipettor. That particular pipettor had been used to spike experimental samples with high-activity carbon-14, needed to trace the isotope into not just the phytoplankton cells, but part of their intracellular components, their pigments. Over time, some isotope had lodged in the barrel of the pipettor, giving spurious results for contamination checks. For experimental use, this level of contamination would not be noticed against the high levels used in Don's experiments. But that's not the case when checking against background radiation on the ship. We were all relieved, but the owner of the contaminated pipette looked a bit embarrassed and sheepish. To console him, we bought him 25-cent beers from the ship's vending machine. Most importantly, the ship was still "clean."

There was yet another application of carbon-14 to the bag experiments. On another cruise, this time to the Sargasso Sea, we spiked one of Curt's bags with bicarbonate labeled with carbon-14, as usual. We wanted to know at any point what the total radioactivity was, to be able to calculate rates of photosynthetic carbon assimilation. We took the initial, or "time-zero," sample, as usual, and when we counted the sample on the ship's liquid scintillation counter, we got about the right amount of activity, given the volume of the bag and how much we had added. However, when we took a subsequent sample, a few hours later, the total activity had declined. It declined again in a sample taken still later. The total activity is not supposed to decline. The phytoplankton have a negligible effect on the amount of carbon-14 added to a seawater sample. When I plotted the results, the total radioactivity in the bag looked to be declining exponentially. No, the seawater in the bag was not aging before our eyes. There had to be a leak. The water in the bag was exchanging with the ocean water outside, a loss at a constant percentage of the volume of the bag. At that point the experiment became untenable, and it was called off. But we had discovered an unexpected use of a tracer: testing the integrity of an experimental vessel, the big bag.

In 1985, communicating with friends or loved ones was not easy, even on a major oceanographic vessel like the *Melville*. Today, that seems like ancient history, only 30 years ago. Sunday afternoons at sea, a lot of us were lined up for a short phone call home—although "phone call" stretches the concept. The ship let us use its single-sideband radio to contact a ham operator somewhere in the English-speaking world, who gave us the courtesy of relaying our voices to the phone system. The sounds were funnily distorted, but communications home were achieved. Somehow, on one of those calls, I must have gotten word to a colleague at Lamont-Doherty, saying, in an offhand way, that our cruise results doubled ocean productivity. That message found its way to the Ocean Sciences Program at the National Science Foundation. We had created some "buzz" in the oceanographic community even before we returned home.

Doubling ocean productivity was about right for the major finding on the PRPOOS cruise. If we were able to "solve" the problem of ocean fertility in an ocean desert, where the differences were most acute, and which occupied a substantial fraction of the ocean as a whole, then we might be safe in extending the result from one cruise to the rest of the ocean. We arrived at what can be called a canonical value for a rate of primary photosynthetic production in the ocean, about half a gram of carbon fixed under each square meter of ocean surface each day.

The PRPOOS cruise was a single occurrence: a snapshot of ocean conditions that might have prevailed at the time but might not happen even a few months later. Three years after the PRPOOS cruise, however, the University of Hawaii initiated the "Hawaii Ocean Time-series," or HOT.[2] The monthly HOT cruises, to a location not far from the PRPOOS site the *Melville* had occupied, also made measurements of primary production based on carbon-14, and the results agreed with those of PRPOOS. That number, about half a gram of carbon fixed under each square meter of ocean surface each day, has survived the test of time and become the typical value to which other results from other locations and times can be compared.

The controversy about ocean fertility still simmers, keeping everyone searching for new ways to assess it. But now I can revisit the older, fiercer, controversy.

The ocean "desert" might explain the differences in fertility reached by Riley and Steemann Nielsen back in the 1940s and 1950s. In actuality, although they probably did not realize it at the time, they sampled two different ocean regimes. Draw a line, mentally, from Halifax, Nova Scotia, due south toward Bermuda, and from about 33°N, draw another line west to Cape Hatteras (see figure 11.1). That area of the Northwest Atlantic had been the backyard swimming pool of most of North American oceanography. Even into the 1970s, the Bedford Institute of Oceanography in Nova Scotia, the second largest oceanographic institution in the world (after Scripps in La Jolla, California), had run quarterly cruises from Halifax Harbour to Bermuda and back. Woods Hole Oceanographic Institution similarly focused on this sector of the Atlantic Ocean, and it was where Riley, Stommel, and Bumpus derived their ideas about plankton ecology, as I mentioned in chapter 11. Within a few days, ships could travel from waters overlying the continental shelves along New England and New York to the "slope water" beyond the shelves and over the continental slope, followed by the Gulf Stream, and finally, the Sargasso Sea (figure 12.2). In winter and early spring, you would be bundled up in a parka and watch

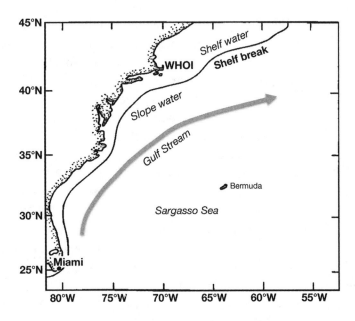

FIGURE 12.2

The Northwest Atlantic Ocean, showing continental shelf water, slope water (over the continental slope), the approximate position of the Gulf Stream, and the Sargasso Sea. For reference, the locations of the Woods Hole Oceanographic Institution (WHOI) and Miami are also indicated, as is Bermuda.

cap leaving Woods Hole or Halifax and quickly shed coats, foul-weather gear, and gloves, crossing the warm Gulf Stream and entering the balmy waters of the Sargasso.

The weather in this part of the Atlantic comes from the west. In winter, winds blow off the North American continent, cooling the ocean, forcing it to give up heat. Cooling the surface ocean allows the waters of the northern areas of the Sargasso Sea to mix over depth, homogenizing surface waters with those from the deep, and thereby bringing deep-water nutrients to the surface where the phytoplankton can use them. The winds in winter begin a seasonal cycle and, consequently, a seasonal cycle of productivity—modest by comparison with other regions, but observable.

South of Bermuda, the effects of the continental winds disappear, and a different, calmer ocean prevails. The seasonal cycle largely disappears.

The Sargasso Sea south of Bermuda has some of the clearest waters on the planet. Clear water, of course, means that there is nothing in it, and therefore few phytoplankton to photosynthesize. These waters and those farther east and away from continental influences were those sampled by Steemann Nielsen during the *Galathea* Expedition. Riley and his coworkers confined themselves to the more seasonal ocean to the northwest.

A few months before the PRPOOS cruise, in the early spring of 1985, I completed a different cruise to the Sargasso Sea, the first trip of an ONR-sponsored Accelerated Research Initiative called Biowatt, a moniker given for the official name of "Bio-optical and Bioluminescence Variability in the Sea." We incubated samples with carbon-14 (added as bicarbonate) in the ocean, attached to a small buoy floating free from the ship, similar to the PRPOOS measurements I did a few months later.

These experiments were done somewhat differently than Riley's oxygen-change measurements, in which samples were incubated on the deck of the ship, but the results we got were closer to his numbers than to Steemann Nielsen's. Soon after the Biowatt cruise, I wrote to Riley about our results. He died in October that year. His wife, Lucy, wrote me after his death and said Gordon was very grateful to hear about the more recent data, apparently relieving him of a long-carried burden. So, differences in where in the ocean samples were collected were important, and could color one's perspective. Steemann Nielsen saw the open ocean as the tropics, maybe unchanging. Gordon Riley saw a more dynamic ocean from the changing seasons—a more weather-dependent ocean.

Another difference has to do with how the ocean is studied, biologically. A tension has always existed between those who enclose small volumes of water for study, as in the carbon-14 method, and those who try to discern ocean dynamics and productivity from the water properties (and the organisms) themselves. I alluded to this dichotomy in chapter 11. The ocean is big; plankton are small to microscopic. Do we enclose small volumes of those microscopic organisms, remove them from their habitat, and observe what they do under these unnatural conditions? Is this at all representative?

The other approach, however, might not be representative either. Studying the evolving ways of the ocean to derive productivity means working with a very changeable, movable environment. Often, these methods involve estimating the state of the ocean in its top layer, which, because it is affected most strongly by winds and the sun, continually changes. How do we account for how the ocean moves and swirls, not just day to day, but hour to hour? "The ocean is motion," the old saying goes. How do we discriminate changes in water properties from changes in the depth to which the ocean can be affected by the wind? Even when the ocean is quiescent, estimates based on water column properties are usually averaged over many days, or a week's time. In many instances, we do not understand the dynamics of the ocean at the day-to-day timescale to be able to get good estimates of productivity, which changes not only day to day, but hour to hour.

The water-column method has the advantage of creating a budget: how much sunlight comes in, and how much productivity can result. It gives a number that can serve as an important check on the system. Holding water samples in bottles, the incubation method, has the advantage of revealing exactly which organisms are photosynthesizing—diatoms, cyanobacteria, flagellated protists, etc.—and how those producers interact with rest of the ocean's food web. In the end, both methods, both perspectives, are necessary.

Steemann Nielsen made conclusions about ocean productivity based on what happens in a bottle. Riley and his coworkers favored evidence from a broader set of trophic (phytoplankton, zooplankton, etc.) variables, but were limited by how well they could delineate environmental variability. So, who was right about global ocean productivity?

Steemann Nielsen's numbers for the organic production in the ocean, at least in certain areas of the North Atlantic, have not withstood the test of time. Subsequent measurements in the late 1950s through the early 1980s showed values about 10 times higher than reported for the *Galathea* Expedition.[3] The data are difficult to compare and to pin down because the locations are not the same (although similar) and the autotrophic biomass responsible for the primary production might differ (or not be measured at all), plus all the usual methodological differences. The controversy could not be laid to rest.

Current estimates are that Riley's global estimate was about a factor of two too high, for the ocean; Steemann Nielsen's was a factor of two or three too low. Today, we think that the ocean produces about 50 or 60 gigatons (10^9 grams) of carbon every year, in between Riley's and

Steemann Nielsen's numbers. (For comparison, a gigaton of water would occupy about a half-million Olympic-size swimming pools.)

There is a whole other dimension to the problem regarding the fertility of the ocean. Productivity occurs at different levels and with different fates. Energetically, we can speak of gross photosynthesis, or gross photosynthetic production. The gross number has an analog in business, or your tax return, as the income before anything is subtracted. For phytoplankton, the major subtraction is their respiration. If we take that away, we end up with net photosynthetic production, and this is what can be made available to zooplankton and, in turn, to fish. Then, after the heterotrophic respiration is subtracted, what remains is net (ecological) community production.

Permutations and processes that might go unmeasured complicate the whole scheme, and the methods themselves—notably, oxygen changes and the carbon-14 method—actually measure different things. An oxygen change from the start to the end of an experiment can measure net community production directly, because the method measures all respiration—phytoplankton, zooplankton, and bacterial—that occurs in a bottle or happens in the water column. Some assumptions and calculations are required to derive gross primary production, and even more to derive net primary production.

The carbon-14 method measures an assimilation, or accumulation of carbon into particulate matter. A plus is that carbon-14 will only be taken up by photosynthesis. As the label accumulates in the cell, and enters metabolic pathways, some is sure to be respired after a time. This means that whether carbon-14 assimilation measures gross, net primary, or net community production, or even biomass, depends strongly on how long the phytoplankton are exposed to carbon-14—that is, how long the experiment is conducted. The length of an incubation remains a contentious topic. We saw in chapter 6 that Calvin, Benson, and Bassham had to end their experiments after seconds to see the first products in algal photosynthesis. It is likely that carbon-14 assimilated into the phytoplankton cell quickly covers the metabolic distance from the chloroplasts, where photosynthesis happens, to the mitochondria, and then is respired in a relatively short time, meaning that the method estimates net primary production after some minutes to an hour or so.

It is said that the only variable that can be definitely and truly measured in the ocean is the water temperature, although some regard even that as being open to question. Certainly, that other basic ocean property—its salinity—is an operative variable, an agreed-upon construct, because the dissolved ions that make up what we understand as salinity themselves vary. At the other end of the spectrum are measures of ocean fertility, sometimes easy to define, but difficult to quantify and evaluate. You would think that the simple measure of the incorporation of carbon-14 into oceanic particulate matter would likewise be simple to interpret, but such has not been the case.

Twenty years ago, an international program called the Joint Global Ocean Flux Study established protocols for the various "core" measurements that would be made during all the expeditions, and also for two open-ocean sites, each occupied on a monthly basis, year-round, year after year. At least, then, we have a way to compare results, at least approximately, across investigations and over geographic areas.

The use of additions of carbon-14 to seawater samples, followed by its assay in particulate matter, finds application across a wide variety of oceanographic research endeavors. There are now other means to estimate ocean fertility—and specifically, the photosynthesis in the ocean—such as the kinetics of the early stages of photosynthesis, the so-called light reactions (see chapter 6). At the long end of the timescale for ocean fertility, Earth-orbiting satellites can now view the ocean as a whole and record the distribution of the abundances of phytoplankton from the color of the ocean. Whether measuring the initial milliseconds of light absorption in phytoplankton or the longer-term changes in their abundance spatially, carbon-14 is used in interpreting and validating the results. These measures are assessed and evaluated by their correspondence with the carbon-14 method.

PRPOOS represented a turning point in biological oceanography. How can one cruise do this? First, as I mentioned above, PRPOOS investigators went to an area of the ocean where the differences between various assessments of fertility were extreme. If we could figure out the problem in what was known as an ocean desert, we might more easily extrapolate the results to other ocean regimes and conditions.

Second, the PRPOOS program used a variety of methods. For just about the first time, the PRPOOS researchers were able to test the carbon-14 method against other techniques—notably, the venerable Winkler method for measuring changes in the oxygen dissolved in seawater. Peter J. LeB. Williams and his (then) postdoctoral fellow, Duncan Purdie, had looked at every conceivable way, at each step in the procedure, that the Winkler method could be optimized to make it as sensitive as that for carbon-14 assimilation.[4] One important improvement was simply to increase the number of replicate samples for better statistical discrimination of the signal in oxygen changes. Michael Bender and his student Karen Grande,

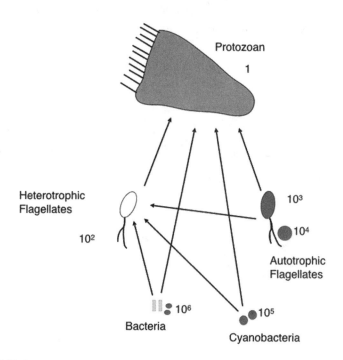

FIGURE 12.3

One schematic of the microbial food web, drawn by Michael Landry. Autotrophic flagellates and cyanobacteria are the photosynthesizers. Heterotrophic flagellates and protozoans consume the autotrophs, and bacteria utilize the waste products. The numbers refer to their approximate abundances in a milliliter of seawater.

then at the University of Rhode Island, introduced an isotopic method using oxygen-18.[5] Oxygen-18, a stable isotope, is added to a sample as $H_2^{18}O$ and then assayed for the dissolved $^{18}O_2$ evolved from photosynthesis after being incubated. The oxygen-18 method was designed to estimate gross photosynthesis and could therefore be compared to the dissolved oxygen fluxes and the assimilation of carbon-14.

Finally, the PRPOOS investigators were prepared to assess the trophic relationships among the microscopic, and not-so-microscopic, plankton—essentially, who was eating whom. One realization, in the midst of the PRPOOS project, came with an appreciation of an entire food web that can occupy a milliliter of seawater. Very small photosynthesizers, together with their very small predators, microscopic protists, and bacteria to consume the leftovers, all existed in that tiny volume (figure 12.3).

13

CARBON-14 AND CLIMATE

The Lamont-Doherty Earth Observatory sits atop the Palisades, the 100-meter-tall cliffs on the west side of the Hudson River in New York. The Palisades form a geological sill dating from the time of the dinosaurs (about 200 million years ago), exposed and modified more recently by the receding glaciers of the last ice age. The Palisades Interstate Park borders the observatory on both the south and north, excepting the small village of Palisades—a post office and a traffic light—and a neighborhood on the banks of the Hudson called Sneden's Landing, which was the original name for Palisades. Lamont-Doherty is three miles from pretty much anything.

Except for a couple of new constructions, Lamont looks almost like a summer camp, with its buildings spread across 100 acres in a woodsy setting. It was the estate of Thomas Lamont, a financier who worked for J. P. Morgan in the early part of the twentieth century. Many of the old Lamont Estate structures still exist: Lamont Hall, a couple of houses that were used by estate caretakers, a greenhouse, an indoor swimming pool—and all have been repurposed to the needs and hospitality of research.[1] One of the early new structures, the Oceanography building, is constructed of blond brick and situated on the Palisades cliff with a breathtaking view of the Hudson River. The side of the building facing the river is painted green to appease the residents of Westchester County, on the other side.

When I arrived at Lamont in the late 1970s, I was only dimly aware of the research on paleoclimates going on, wrapped up as I was in my own at-sea programs and laboratory projects of a more present-day nature.

I told the story of some of that research in the last chapter, on the use of carbon-14 in the study of ocean fertility (productivity). Carbon-14 was also a foremost isotope in climate research, but was orders of magnitude away from me in time, radioactivity, and, in those days, appreciation.

■ ■ ■

Recently, I sat across from Wally Broecker in his office at Lamont, and he proceeded to trash every idea I had entertained myself with about Earth's climate 10,000 to 20,000 years ago, and to criticize pretty much all the existing ideas, his own and everyone else's. At the same time, he listens to all views, accepting competing ideas and hypotheses until evidence proves them untenable.

Wally's office is in a commanding location, on the second floor of the new Comer Geochemistry Building at Lamont-Doherty, overlooking the walkway to the front door. Like a bridge on a ship, it even has a ship's wheel in the window, symbolizing his position in climate science worldwide. Wally's doctoral dissertation treated "radiocarbon" in the ocean. That was in 1957, still early days after the discovery of carbon-14. He has been researching carbon-14's role in climate and in the ocean ever since, for 60-odd years.

Sixty years later, the changes shown by carbon-14 remain a mystery. Carbon-14 values recorded and documented in the ocean and its sediments still stir up puzzles. We're talking about times beyond human history, beyond archeology, beyond a couple of half-lives of carbon-14. In some sense, it is amazing we can find out anything at all from 10,000 to 20,000 years ago, before humans started writing things down, building ceremonial monuments, or smelting metals (see chapter 8). Climate "proxies" allow us to do that, to reconstruct past climates, if only imperfectly or approximately.

Climate proxies are things, artifacts, or indicators of past climates—stand-ins for meteorological measurements that were never taken. They can be the remains of organisms, elemental isotopes, chemical signals, wood, pollen, macrofossils, charcoal. Proxies are central to studies of climate, and their meaning will become clearer as we explore and investigate the past. Carbon-14, although not a climate proxy itself, is a big part of the story, allowing a connection between a climate condition and a time interval.

The pathways of carbon in photosynthesis (chapter 6), the discovery of carbon-14 dating (chapter 4), the "new archeology" using those dates (chapter 8), and understanding ocean circulation and fertility (chapters 9, 11, and 12), all unfolded in the second half of the twentieth century. The story of carbon-14 and climate, on the other hand, continues today. For climate, unlike those other efforts, there was not so much a revolution in the science as incremental steps spread out over the last few decades of the twentieth century and continuing into the twenty-first. Extending the records further back in time changed the perspective of the earlier and more limited chronologies. What might seem like a large signal over 10,000 or 12,000 years of chronology can become much less significant in the larger, or longer, perspective.

Understanding climate changes means considering (at least) the record since the last glaciation. In the 1970s, it became possible to reconstruct the history of the demise of the ice sheets that covered much of the northern hemisphere from ice cores, using annual snow accumulation records (to identify years) together with values for the stable oxygen isotopes (the ratio of oxygen-18 to oxygen-16) to estimate temperatures.[2] The most recent ice age had its maximum about 22,000 years ago. Geologists and climatologists call this the "Last Glacial Maximum," or LGM. About 19,000 years ago, the ice sheets began to melt, receding from much of the landscape.

"Chronologies are critical," Wally tells me from across his desk, regarding the various climate indicators, or proxies. The proxies contain the clues that we have not figured out yet, but ordering them in time is paramount. And as the records recede in time, the chronologies differ, sometimes markedly. One thing that they all agree on is that Earth's climate has undergone change over the past 20,000 years, and that the change was not simply or smoothly from a glacial to an interglacial (a warm interval that we are experiencing now). It happened in fits and starts, and at times the changes occurred abruptly. The various jumps began to be perceived in the last century, by paleobotanists in Scandinavia. That's where the story of climate chronologies begins.

■ ■ ■

Nineteenth-century Scandinavians relied on peat as fuel for their stoves. They might have needed extra fuel to get them a little more comfortably through the "Little Ice Age," a moderately cold climatic period (about 1°C cooler, on average) from around 1350 to about 1850.

Peat was then, and still is, readily available. Mined as bricks, it is what remains of plant matter when it is buried in a wetland and unable to decay completely. Wetland environments are storehouses of organic matter and easily become deficient in oxygen. At cool temperatures, waterlogged, and without oxygen to fuel it, decomposition slows down. The partially decomposed plant matter, peat, accumulates over substantial depths.

Digging up the peat showed old tree and plant debris and also revealed a distinct layering, much like geological formations. The layers and debris within them caught the interest of local botanists. The debris included leaves, notably from an Arctic and high-mountain flowering herb, *Dryas octopetala*. The *Dryas* leaves appeared and disappeared at various depth intervals in the peat layers, intermittent with tree leaves. Since *Dryas* grows in the Arctic or on mountaintops, its presence indicated intervals of a cold climate—tundra. The tree leaves indicated warmer climates, a boreal forest. Later, under the microscope, they identified tundra pollen, and these examinations eventually marked the beginnings of "palynology," the science of fossil pollen as indicators of trees, shrubs, and herbs, and the plants themselves as indicators, proxies, of the environment that existed when they were buried.

In the late nineteenth and first part of the twentieth centuries, Scandinavian botanists developed chronological sequences based on the peat layers, called "chronozones," and correlated these over wide geographic areas, at least in northern Europe. The sequences were all relative in terms of age; deeper pollen associations were older, and shallower ones were considered younger. The layers separated sharply, and were easily identified. *Dryas* episodes were familiar to European palynologists. North American palynologists saw different species changing, and made similar climatic conclusions, although with some disagreement, as the climate proxies or signals in North America were initially not as widely understood.

The *Dryas* episodes that the Scandinavians identified were cold intervals, but the transitions between the episodes (peat layers) were geologic "instants," rapid changes, perhaps within a few decades. According to later carbon-14 dating, pollen and leaves from *D. octopetala* suddenly started appearing about 12,900 years ago and just as suddenly disappeared 11,500 years ago. This climate period, lasting over a millennium, came to be known as the Younger Dryas, for the indicator plant. "Younger" refers to the most recent, shallow, location of the *Dryas* pollen and debris, compared with two other deeper, previous occurrences, which the Swedish botanists identified as the Older and Oldest Dryas periods. These names represent the relative chronology of the layers in the bog, but they persisted even after absolute dates became available with radiometric dating. "Younger Dryas" became, at least for palynologists, geologists, and climatologists, a household name, known globally, even though the *Dryas* chronozones were seldom recognized outside of Europe at the time.[3] The Older and Oldest Dryas periods were much shorter, and were never as intensely studied as their Younger sibling was.

The temperature changes deduced from a Greenland ice core for the past 15,000 or so years are shown in figure 13.1.

Humans also had to adjust to the swings in temperature. Prior to the Younger Dryas, there was a period of warmth called the Bolling/Alleröd—Alleröd being the name of a clay pit in Denmark, and Bolling, a lake, also in Denmark. Environmental conditions during the Bolling/Alleröd were identified from palynology, with greater amounts of pollen from temperate forest species, conifers, and some broadleaf trees. For humans, the wild pickings were probably pretty good during the Bolling/Alleröd, and could support hunter/gatherer populations. Wild stands of grains were abundant enough to be harvested where they were found, and not have to be cultivated.

For that last big change, the Younger Dryas, it was almost as if inhabitants of northern Europe and North America woke up one morning to colder weather that soon became a cold spell that didn't quit. It wasn't long before they began to experience what aboriginal populations of the north called a "three-dog night." Working dogs and humans in those days lived close together. Both needed the extra warmth.[4]

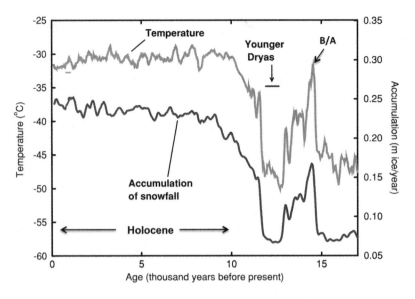

FIGURE 13.1

Temperature changes for the past 17,000 years in central Greenland retrieved from an ice core and calibrated with oxygen-18/oxygen-16 ratios. Accumulation of snowfall (as a rate) is derived from the thickness of annual layers (see note 2). A colder Earth is accompanied by more snow. Shown are the Bolling/Alleröd warm period ("B/A"), the Younger Dryas, and the period of the Holocene (after Committee on Abrupt Climate Change 2002).

The Younger Dryas meant that human populations that were affected had to organize and plan. Hamlets indicative of sedentary population grew more developed, particularly in West Asia, the Levant (present-day Lebanon, Israel, Iraq, Syria, Turkey), but also in East Asia. Humans found that they had to work harder, grow their own food, cultivate grains, and as a result, settle down. In part because of its timing, the Younger Dryas, the last major shudder in the climatic shifts during the long-term retreat of the glaciers, is thought to have had long-lasting consequences for human history. There were exceptions to this general trend, but overall, the Younger Dryas seems to have meant more sedentary human populations.[5]

The abruptness and the putative large temperature change occurring with the Younger Dryas called out for an explanation. The Younger Dryas became famous among climate scientists and paleoclimatologists. It was, in the 1970s and 1980s, a hot topic. Also in the 1970s, larger-scale climate change itself came under a new, or rediscovered, theory—that of variations in Earth's orbital cycles, formulated by the Serbian geophysicist and astronomer Milutin Milankovitch.

Earth does not spin around perfectly on its axis, but, like a spinning toy top slowing down, it wobbles. Also changing is the angle of tilt of Earth's north-south axis, the tilt that gives us our seasons. Earth, currently at 23.45°, can become more or less tilted, from about 22° to 24.5°. Third, Earth's orbit around the sun is not quite circular, but elliptical. The gravitational pull of other planets changes so that the path the Earth traces as it orbits the sun varies slightly, from more to less circular, less to more elliptical. All these variations—the wobble, the tilt, and orbital variations—happen over different time intervals, ranging from 22,000 to 100,000 years. The deviations mean that solar radiation reaching Earth varies—not by very much, but the effects accumulate. A little less sunlight arriving at the Earth in the spring months means slightly colder temperatures, more snow, and a little less snow melt. Less snow melt then contributes to a little greater solar reflection off the snow, feeding back to overall colder temperatures, and so on. Pretty soon (on geological timescales) you have an ice age.[6]

Milankovitch theory helped explain the gradual alterations to climate— the changes from glacial to interglacial climates—that happen over hundreds of thousands of years. But there were two concerns or issues. The first was whether the small changes in solar radiation by themselves were sufficient to cause the large changes to climate. Second, there were shorter-term, more abrupt climate shifts—for example, the Younger Dryas—that happened on top of the gradual changes predicted by Milankovitch theory.

A review by Broecker, coauthored with Dorothy Peteet and David Rind, both working at NASA's Goddard Institute for Space Studies in New York City, looked at the first issue: whether small but additive changes to solar radiation were enough to create either glacial or interglacial climates. Broecker, Peteet, and Rind (1985) found good evidence for the transition between glacial and interglacial conditions over Milankovitch timescales—but they also noted many "brief events" that could not be dismissed as "climate 'noise.'" For their analysis, the team used changes to CO_2 and the ratio of oxygen-18/oxygen-16 found in bubbles inside cores of ice drilled and retrieved in Greenland (see figure 13.1). The CO_2 represents atmospheric concentrations over Earth's past from recent times back to the Last Glacial Maximum (LGM), 22,000 years ago.

The oxygen-18/oxygen-16 ratios indicate source water values, water transport, and temperatures. The derived temperatures in the ice-core records that Broecker and his team analyzed correlate with concentrations of CO_2. When CO_2 is low, so are temperatures, and vice versa. After considering various possibilities for the rapid variations to CO_2 and temperature in the atmosphere—oceanic mixing, biological effects, and a much lower sea level during the LGM—Broecker and his coauthors concluded that the changes between glacial and interglacial climates have a mechanism in ocean circulation. The ocean contributed to the changes in climate, from glacial to interglacial.

During the interglacials (preindustrial warm conditions), seawater descending to depth in the North Atlantic forms North Atlantic Deep Water (NADW) and enters into the global conveyor, as I discussed in chapter 9.[7] Broecker, Peteet, and Rind argue that during glacial times, the formation of water sinking in the North Atlantic must have weakened, been reduced, or might even have been absent. They use other kinds of evidence—notably, a proxy for the amount of phosphate in the water that is preserved in the sediments, in the shells of foraminifera (discussed later in this chapter). Higher phosphate in the shells from glacial times indicates that deep water did not form.[8]

Broecker and his colleagues' ideas contributed to mechanisms for glacial-interglacial changes in climate, but what about the shorter-term, abrupt changes such as the Younger Dryas? When they published their review, the cold snap of the Younger Dryas was thought to be a regional phenomenon, confined to the land areas surrounding the North Atlantic, largely in Europe. Even though evidence outside of Europe—for example, in Atlantic Canada—remained sparse, the numbers for an ocean influence seemed to add up, and suggested a more global phenomenon.

Deep-water formation in the North Atlantic supplies a transport of 20,000,000 cubic meters per second,[9] about 20 times the combined flow of all the rivers in the world entering the ocean—the Nile, the Amazon, the Mississippi, the Congo, and all the rest. As we saw in chapter 9, the formation of NADW means a loss of heat to the atmosphere, and that atmosphere warms up the areas downstream of the North Atlantic Current: Iceland, the British Isles, Scandinavia, and the rest of northern Europe. The amount of heat released represents about 30 percent of solar heat at the latitudes where deep water is formed.

When deep-water formation is turned on, therefore, there is a source of heat to the atmosphere, and the atmosphere can warm. When it shuts down, the atmosphere cools, or remains cool. Here is a mechanism that is not based on the gradualism of Milankovitch cycles, but is tied to ocean circulation. Deep-ocean circulation changed between glacial and interglacial conditions, changing over thousands of years, as I have discussed. The next question is: Can ocean-circulation changes happen fast? Can they explain the sudden changes in climate, of a few decades, ushered in by the Younger Dryas?

In chapter 9, again, we learned that deep-water formation in the northern North Atlantic relies on warm, salty water from the south, from the Gulf Stream and North Atlantic Current. Another way to inhibit sinking in the North Atlantic is to change the salinity. Adding fresh water—for example, from melting glaciers or river inflows—reduces the salinity, making the waters less dense.

The receding glaciers in central North America became a more or less obvious early candidate for shutting down the deep-water formation. The melting ice sheet created a huge, postglacial lake covering most of what is now Manitoba and much of western Ontario—a massive area, about 170,000 square miles. Even though it was not deep, that is still a lot of fresh water, about as extensive as the Black Sea. The lake was named Lake Agassiz (figure 13.2) after the nineteenth-century naturalist Louis Agassiz. With an elevation of almost 300 meters (1,000 feet) above sea level, Lake Agassiz could only drain into the Mississippi River to the south, prevented as it was by glacial ice to the north, and ice to the east in the Great Lakes.

The scenario has it that an ice dam burst catastrophically, allowing Lake Agassiz to empty, and quickly, to the east, with its fresh water cascading down elevation, successively through Lakes Superior, Huron, Erie, and Ontario, and into the St. Lawrence River valley, ultimately flooding the North Atlantic with fresh water. The flood of fresh water slowed or reduced deep-water formation and thereby rapidly cooled the northern hemisphere, reverting its climate to a colder state, the Younger Dryas.[10]

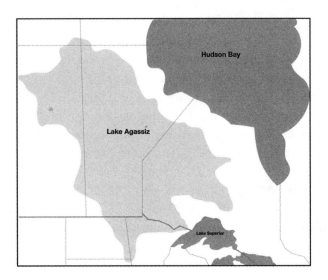

FIGURE 13.2

A rough outline of postglacial Lake Agassiz. The two possibilities for draining the lake are (1) eastward through Lake Superior (followed by Lake Huron, Lake Erie, Lake Ontario, and the St. Lawrence River Valley) or (2) to the northwest, through the Northwest Territories, the MacKenzie River, and the Canadian Arctic. See Broecker 2003.

All these ideas came about before carbon-14 data in climate proxies became readily available. Progress in the research required measuring natural abundances of carbon-14 in small samples, and in the early to mid-1980s, accelerator mass spectrometry (AMS) methods (see chapter 7) were still being worked out. Explaining the sudden changes of the Bolling/Alleröd and Younger Dryas episodes relied instead on the various proxies by themselves. The Younger Dryas, at only two to three half-lives in the past (10,000–12,000 years), seemed an ideal application for carbon-14. It was near enough in time for good radiometric dates with carbon-14, and the proxy signals for the environment were large enough and occurred over a long enough time to delineate past climate changes—a chronology. A further motivation was that making sense of past climate changes could also help figure out what humanity might expect to see in the near future. The dilemma, as we will see, is that even though near in time, geologically speaking, postglacial changes still remain far away in understanding.

■ ■ ■

By the 1970s, a tree-ring chronology had been established for carbon-14, pushing verifiable dates for atmospheric carbon-14 back to just about the Younger Dryas, about 13,000 years ago. Only recently have dates been retrieved earlier than 12,000–14,000 years before the present. Beyond that time, carbon-14 dates have had to be retrieved from various other proxies. I'll now review these proxies and methods for what they say about climates in Earth's past since the last deglaciation. The proxies are in trees, single-celled protists, corals, caves, and oxygen-deprived environments, similar to the peat bogs from Scandinavia. One has to recognize that ocean-atmosphere exchange, ocean mixing, and upwelling will also influence these proxies. They may have even been affected by conditions of the Younger Dryas itself. In many instances, it is difficult to know if one is measuring an age or an Earth process.

That tree stumps have concentric rings when viewed in cross-section has been known at least since the ancient Greeks, and no doubt by many other earlier, undocumented, observers. During the Renaissance, Leonardo da Vinci reported that the rings were formed annually, and French scientists noticed that variations in the width of the rings aligned with notable weather events—a key finding for establishing chronologies. Charles Babbage, of nineteenth-century computer fame, realized that the growth rings might be used to date the remains of trees found in peat bogs.

Trees chronicle the climate in which they grow, and especially in temperate climates, they record seasonal changes. Bristlecone pines, as mentioned in chapter 4, grow very slowly, and live for upwards of 4,000 years. The Centennial sequoia had about 2,900 rings. Otherwise, the vast majority of trees do not live so long and are not usually cut down, like the Centennial. Trees, however, can be cored, sort of like taking a plug of a watermelon to check ripeness. Tree cores, of course, have a multiplicity of layers instead of just the skin, rind, and pulp.

What scientists do to get around the shorter lifetimes of trees with respect to the longer "lifetime" of climate, is to create groups with different age ranges. They then match the different groups of trees based on a common stress year—for example, a year of drought, or a very cold winter when little or no growth occurs. By matching groups of trees, and overlapping and matching the rings, scientists can create a longer-term chronology—or, since these are trees, a dendrochronology. The scientists who make these analyses are called dendrochronologists. Tree-ring analysis can take the climate back pretty well to about 10,000 years, and perhaps even to 14,000 years, just beyond the Younger Dryas. To get further back requires record keeping of a different sort, and a different environment.

One of the "different sort" of record keepers, in a different environment, are foraminifera. I introduced foraminifera ("forams") in chapter 10, as one endpoint to the ocean's carbon cycle for carbon-14, their shells becoming buried in the ocean's sediments. Forams are single-celled protists, and throughout their short life, they create chambers arranged in a spiral and made from calcium carbonate (figure 13.3). The "carbonate" means they also incorporate carbon-14. Forams live for a few weeks or a month in the ocean's surface layers before they die, and their shells, about a millimeter or two in

FIGURE 13.3

The foram genus *Globerigina*, about a half-millimeter across in size.

size (like a typical grain of sand), sink to the abyss. Their short span in life and their potentially long burial in the sediments make them ideal record keepers of the conditions at the ocean surface at time points in the past.

Other forams spend all their time at the bottom crawling over the ocean sediments, but like their planktonic brethren, they too die and get buried. The sediments where the forams are buried, now "fossils," or "microfossils," can show time horizons, with deeper meaning older. Having carbon-14 in their shells also means they can be dated.[11] Their value as a climate proxy, however, comes from the fact that different species of forams favor different temperatures. Finding a particular planktonic species at a particular depth in the sediment associates that age with a particular temperature of the seawater (usually, warm or cold) in which it lived.

Climate record keeping using forams has a few requirements. First, the sediments where the forams end up cannot be too deep in the ocean. Deeper than about 3,000 meters (although this varies), the water becomes corrosive to the shells, and they dissolve.[12] Second, the rain rate of particles from the surface has to be relatively rapid. If the sedimentation rate from the surface is too slow, bottom animals eat, excrete, and otherwise mix the sediments where one had hoped to be able to discern distinct, age-related layering, and therefore a clean record. Finally, if the sediments and forams are deposited near a continental margin, shallower sediments may slump, plummet to deeper depths, and bury younger sediments with older, or at least differently aged, ones.

Planktonic forams record the conditions near the ocean's surface—for example, the carbon-14 that has recently entered the ocean from the atmosphere. Living their lives at the ocean bottom, benthic forams record the amounts of carbon-14 in the deep sea. The difference between the two amounts, shallow and deep, measures how long, in carbon-14 years, the waters of the deep sea have been isolated relative to the surface. For example, benthic forams in the Pacific have carbon-14 values indicating they are about 1,600 years old, whereas surface forams have dates of about 150 years old.

Although subject to myriad complications, foraminifera shells, as fossils, can push the carbon-14 chronology back to more than 20,000 years before the present, to when the ice sheets from the last glaciation began to melt and recede.

Corals are group of animals that are colonial, building sometimes immense calcium carbonate structures, coral reefs. Like trees, they add layers of calcium carbonate according to seasons, food supply, water clarity, and temperature. They calcify at a rate of millimeters to a centimeter or two per year. The coral skeletal layers vary in density, with a high-density and low-density band added each year. The coral animal itself, similar to the familiar *Hydra* from high school biology labs, houses microscopic autotrophic protists, called zooxanthellae, which photosynthesize. The relationship is close, intracellular, and defined as an endosymbiosis. The coral gets food from zooxanthellae photosynthesis, and the zooxanthellae in turn get a home and a nutrient supply. The corals depend on photosynthesis by their zooxanthellae, and so, needing light, they live and grow near the ocean's surface. That means corals will incorporate carbon-14 into their skeletons representative of the atmosphere as they grow.

To use corals as climate proxies, researchers use high-powered drills to bore into the coral heads and extract a core, layered according to calcite buildup. In 1988, Rick Fairbanks, also a researcher at Lamont-Doherty, found a way to drill into submerged "fossil" coral reefs off the south coast of Barbados (Fairbanks et al. 2005). His cores could also be dated by other radioisotopes,[13] so he and his team pushed the carbon-14 chronology to the Last Glacial Maximum, and later to 50,000 years before the present, closer to what is thought to be the ultimate limit for carbon-14 dating.[14]

As I mentioned in chapter 4 on carbon-14 dating, limestone deposits occur in caves. These deposits are called speleothems, the collective name for the more common stalagmites and stalactites. Groundwater percolates through the soil and bedrock, and this groundwater contains dissolved carbon dioxide from microbial respiration. The carbon dioxide reacts with the water to form carbonic acid (H_2CO_3). The acidic water dissolves calcium carbonate in the rock, and when it enters the cave, the water evaporates. Evaporation leaves behind the stuff dissolved in the water, such as calcium carbonate. Water, drip by drip from above, lands on the cave floor, which when it evaporates, coats a spot there. The spot, over time, grows into a tower, a stalagmite. If the water evaporates before falling to the cave floor, it sculpts a stalactite hanging from the cave ceiling.

Speleothems glisten, showing the evaporation of groundwater, and the amount of groundwater evaporating is a measure of the climate at the

Earth's surface. Thick layers represent wet conditions, and thin layers, drought. The carbon dioxide precipitated in the speleothem represents perhaps a one- to ten-year time window of plant and bacterial respiration in the soil above. Through all this, a small percentage of the carbon is carbon-14. Once collected, the speleothems are sawed along their length, exposing the longitudinal layers. The halved speleothems are scrutinized under a microscope, and small samples of the layers are drilled out for isotopic analysis.

The most complete carbon-14 history comes from speleothems from Hulu Cave in eastern China, near Nanjing (Southon et al. 2012). The speleothems from Hulu were originally used to detect changes to the Asian monsoons, but have proved useful for overall climate reconstructions and chronologies. The data interval for Hulu covers from more than 20,000 years ago to 10,700 years ago, a period that nicely dovetails with the record from tree rings, which go back to 14,000 years before the present (ybp).

Lake Suigetsu is a mixed salt- and freshwater lake on the west coast of Honshu, the main island of Japan. It is one of several so-called "kettle" lakes formed from volcanic eruptions. Suigetsu translates romantically to "water moon lake," even though it is partially anoxic, and at times malodorous. The near-bottom water in the middle portions of the lake is where it is anoxic: there is no oxygen dissolved in the water, and therefore no biology—no organisms, no bacteria, nothing to interfere with the slow accumulation of sediment from above. As a result, Suigetsu has extensive and undisturbed sediment layers called "varves." Sediment cores show individual annual layers visible to the eye, sometimes only one millimeter thick. Even so, varves, or even the cores, individually, are incomplete, especially in the deeper (older) sections. Over the years, multiple cores have been taken and analyzed by a variety of investigators to create a full history. The composite collection of cores show that Suigetsu sediments go back almost 50,000 varve years, extending the carbon-14 chronology back as well. Carbon-14 analyses are done on leaves, small twigs, and insect wings found in the varves. The carbon-14 dates can be calibrated by comparison with other isotopes, as for the corals, but also against known volcanic eruptions, which produce easily identified ash layers.

One of the best records that has been obtained for carbon-14 across the Younger Dryas comes from a core in the Cariaco Basin, in the Caribbean just north of Venezuela. The Cariaco Basin has, like Lake Suigetsu, an absence of oxygen near the bottom. It is a naturally occurring anoxic zone—a rarity in the ocean—a depression surrounded by shallower sills and islands, and therefore isolated from the circulation in the Caribbean Sea above and beyond the sills. The sediments in the depressions quiescently accumulate, without meddling by bottom-dwelling organisms. To ecologists that study the ocean-bottom ecosystem, the sediments are not "bioturbated"—that is, not biologically disturbed. The lack of oxygen, and therefore the absence of biological activity, means a fairly clean record of sedimentation, with easily identified varves—an ocean sediment column that can be clearly dated.

This part of the Caribbean experiences two weather patterns annually, wet and dry, and these conditions are faithfully recorded in the Cariaco Basin sediments (Hughen et al. 1996). Wet periods show a dark-colored sediment and indicate a source from runoff—that is, from the land. Dry periods have sediment with a lighter shade, with microfossils (diatoms, mostly) characteristic of upwelling and higher winds, indicating an oceanic origin. The advantage of the Cariaco Basin over Lake Suigetsu is that it is in the Atlantic, and presumably closer to alterations in ocean circulation.[15]

■ ■ ■

All of these inputs—from tree rings, forams, corals, anoxic environments, and speleothems—have been consolidated into a carbon-14 chronology going back some 40,000 to 50,000 years (see, e. g., Hain et al. 2014). Here, however, I will consider the changes since the beginning of deglaciation 20,000 years ago. For carbon-14, that time span holds the most interest, the most change, and the most puzzles. Beyond 20,000 years before the present, in glacial times, the carbon-14 to carbon-12 ratio does not change much except from factors that change the supply of carbon-14 to the upper atmosphere, as I discussed in chapter 4. However, there is one interesting bump at 40,000 years ago. At that time, the supply of carbon-14 to the atmosphere increased because Earth's magnetic field declined substantially, allowing more cosmic rays and more carbon-14 production in the upper atmosphere. The magnetic pole might have even flipped 40,000 years ago, but for some reason did not.

As mentioned previously, in the late 1980s, a detailed carbon-14 chronology extended only as far as the Younger Dryas, and it seemed at the time like a big signal. As I discussed earlier in this chapter, one thing that most people agreed on is that the cold snap that was the Younger Dryas probably resulted from changes to ocean circulation. Changes to the global ocean conveyor drove the climate, at least in the northern hemisphere. If that was the case, and if carbon-14 is closely related to ocean circulation, it follows that carbon-14 could be used to illuminate what went on 10,000 or 12,000 years ago.

But what caused the Younger Dryas event in the first place? What was needed was a way to reduce or shut down deep-water formation in the North Atlantic. The favored explanation, as I discussed earlier, was that the receding ice sheets somehow left the North Atlantic with substantially more fresh water, which lowered the salinity, and therefore the density, of the surface waters. The reduced density at the surface stabilized the ocean over depth and inhibited or prevented sinking and deep-water formation. The northern North Atlantic appears to be sensitive to factors that promote or inhibit the formation of North Atlantic Deep Water, and therefore it is a prime location for regulating Earth's climate.

The Younger Dryas is easily identified in the Cariaco cores as lighter on a gray scale, and correlated with ice cores from Greenland, where the Younger Dryas has been identified through a change in oxygen-18/oxygen-16 ratios (figure 13.4).

Carbon-14 relative to carbon-12 ($\Delta^{14}C/C$) increased at the beginning of the Younger Dryas. That initial increase in the $\Delta^{14}C/C$ ratio can be explained by the increase in ice cover from the colder conditions, near-glacial air temperatures, and the shutting down of deep-water formation. The reduction, or turning off, of the North Atlantic Deep Water formation at the onset of the Younger Dryas means that carbon-14 forming naturally, or cosmogenically, in the upper atmosphere had less chance of being sequestered in the deep ocean. It therefore accumulated in the atmosphere. In the Cariaco record, the increase in the $\Delta^{14}C/C$ ratio is for the planktonic foraminifera, thought to be recording the surface ocean, and therefore atmospheric concentrations. The increase in the atmospheric ratio goes on for about 200 years—but instead of remaining high during the Younger Dryas, it begins to decline again. Why? Did NADW formation restart?

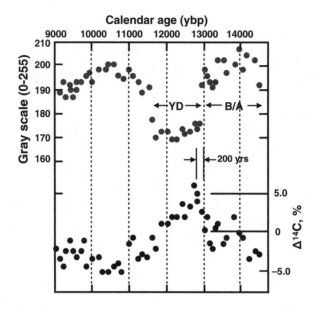

FIGURE 13.4

A sediment record from the Cariaco Basin showing the Younger Dryas (YD) cold episode as a lighter shade of gray, indicating dry, windy, and upwelling conditions, and earlier Bolling/Allerőd (B/A) period. The gray scale, identifying the dates for the Younger Dryas, has been verified from oxygen-18/oxygen-16 ratios from an ice core retrieved from the Greenland ice cap. Also shown is a record of carbon-14/carbon-12, $\Delta^{14}C/C$ ratio (as a percentage change) and its initial 200-yr increase at the YD onset, from foraminifera in the core (after Broecker 2005, from Hughen et al. 2000).

After 200 years, other parts of the ocean must have come into play. The most likely candidate is near Antarctica, in the Southern Ocean, which, through its generalized upwelling and strong mixing, allows a connection between its surface and the deep sea. After the initial increase, over the succeeding 800-plus years of the Younger Dryas, carbon-14 as a proportion of carbon dioxide in the atmosphere slowly declined, as deep water, forming somewhere, but probably in the Antarctic, allowed its penetration into the deep sea.

The Younger Dryas was a sort of "mini ice age," and we can use that idea to help understand what happened and what drove the changes. Here's what we know, and how carbon-14 has contributed to the story.

First, glacial periods have reduced or absent North Atlantic Deep Water formation. The Younger Dryas is an instructive "mini" version of glacial conditions. Second, what happens in the North Atlantic connects to other places globally. The Younger Dryas can be seen not only in Europe, but also on both eastern and western coasts of North America and in the Midwest, as well as in the tropics, such as the Cariaco Basin, in anoxic lakes in Japan, and in caves in China. Third, there is a global dynamic to ocean circulation that operates on timescales of hundreds of years. A change to that dynamic in one place can mean a compensating dynamic in another.

Models of ocean circulation suggest that the meridional overturning circulation (MOC), which I discussed in chapter 9, had a very different character during the last glaciation. North Atlantic Deep Water continued to be formed, but because the North Atlantic was cut off from a supply of frigid water from the Arctic, the surface waters were not dense enough to sink to the deep. Furthermore, the presence of sea ice pushed the formation of NADW to more southerly latitudes.

Models say that ocean and atmosphere dynamics will always force deep-water formation somewhere. Figure 13.5 summarizes these changes, using data from the Younger Dryas episode and the output from ocean models.

In the 1980s, the Younger Dryas, at 13,000 years or so, was as far back as the chronologies could take us. When carbon-14 data from accelerator mass spectrometry became available in the late 1980s to address the cold snap, the changes evident in the Younger Dryas loomed large, generating ideas based on that chronology. By the mid-1990s, however, thanks to speleothems, corals, anoxic sediments with organic matter, and foraminifera, the chronology could be extended back to the limits of carbon-14 dating, 40,000 or 50,000 years. The more interesting piece of the chronology, in terms of carbon-14, is what happened when the ice started to melt—the deglaciation that began about 20,000 years ago. Looking at the big picture changes the perspective. The initial deglaciation is a huge signal in carbon-14. But a signal of what?

Deglaciation, overall, shows a steady decline in the inventory of carbon-14 as a proportion of carbon dioxide in the atmosphere (figure 13.6). We expect the carbon-14/carbon-12 ratio to be higher

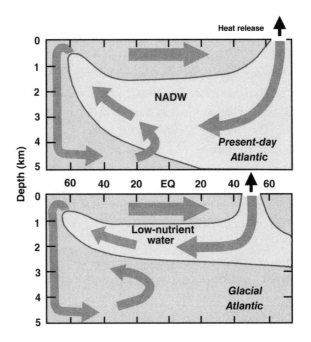

FIGURE 13.5

Cartoon of meridional overturning circulation schemes for the present-day Atlantic Ocean and for the Glacial Atlantic. Vertical arrows indicate where oceanic heat is released to the atmosphere (after Broecker 2005).

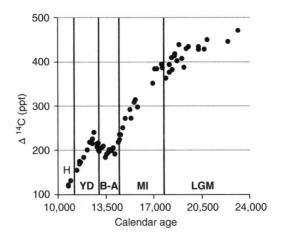

FIGURE 13.6

Carbon-14/carbon-12 ratio ($\Delta^{14}C/C$) chronology in the atmosphere from coral and speleothems. YD is the Younger Dryas, H is the Holocene (the current geological age), B-A is the Bolling/Alleröd warm period, MI is the Mystery Interval, and LGM refers to the Last Glacial Maximum (after Broecker 2009).

during glacial times, when exchange with the surface ocean will be reduced because of extensive ice cover. Models suggest that the glacial ocean also had a modified meridional overturning circulation such that the NADW formed, but the D for "deep" could probably be dropped.

On the timescale since the last deglaciation, the Younger Dryas and Bolling/Alleröd events, on which so much scientific capital was spent, seem less significant—minor hiccups in the overall change. Not that the climates during those periods were unimportant, but carbon-14 was not much affected. The biggest change in carbon-14 occurred during what Wally Broecker has called the "Mystery Interval" between 17,500 and 14,500 ybp. What caused the decline? If carbon-14 is produced more or less constantly in the stratosphere, where did it go?

Earnest attempts at finding the "missing" carbon-14 have been made, looking at deep water zones. Perhaps the increased ice cover over the Antarctic kept the carbon from being exchanged with the atmosphere? Broecker even hinted that if the half-life of carbon-14 were increased, much of the decline be would be eliminated. But that would be a systematic error that would only work for the complete chronology going back 40,000 ybp. The Mystery Interval remains just that, a mystery, although recent models suggest that the source of the decline is in the production of carbon-14 itself, and not in its mysterious sequestration in the ocean somewhere.[16]

Just as mysterious, still, is the Younger Dryas cold snap. Breaking an ice dam, or other such catastrophic flooding, would change things pretty quickly and lead to a long-lasting cold spell. But there also has to be a mechanism for the cold spell to end—for the warming that ended the Younger Dryas as fast as it began, inaugurating the Holocene and our familiar, warmer, less changeable climate. The answer may be that the NADW formation "on" mode of ocean circulation is more stable, and the cold period the aberration—that NADW-on is what the system defaults to, to borrow a phrase from computer science.[17]

That a global circulation cell or global conveyor belt—the MOC— operating as it does now, with substantial deep-water formation in the North Atlantic, is the more stable of two circulation modes reassures. However, the ocean is warming, Arctic ice is melting, the Greenland ice cap is receding—all indicators of a warming Earth. The freshening of the North Atlantic from the melting ice could stabilize the waters there, shutting down the MOC. All these, however, are gradual changes, occurring over decades and longer. At least, catastrophic flooding of the North Atlantic that initiated the Younger Dryas does not seem to be in the cards. Nevertheless, the current climate (Earth's average temperature) also seems unprecedented over the past 400,000 years. We are in unfamiliar climate territory—"uncharted waters."

In the end, as has often been said, predictions are hard to make, especially about the future. One thing is clear: carbon-14 will continue to be central to efforts to uncover the past to inform what lies ahead.

EPILOGUE

The remarkable discovery of a long-lived isotope for carbon, carbon-14, ushered in revolutions in science. Carbon itself has unique properties. It forms the basis of life on our planet, if not the universe. Thus, in 1940, was born a means for greater understanding of life in the present— and life in the past. Life, since its inception, has changed Earth irrevocably. As I pointed out in chapter 2, "carbon-14 is the translator for all that carbon can be: the tag, the message, for what organic molecules are, their history, and how they act."

In the second half of the twentieth century, carbon-14 was exploited in two different ways, leading to two parallel historical threads: one, adding it as a tracer to biological experiments, and the other, utilizing its natural abundance to understand Earth's environments and civilization in the present and the past.

After the interruption of World War II, additions of carbon-14 in carefully designed experiments led to the working out of the cycle of photosynthetic carbon assimilation in algae, and by extension in all plants. Carbon-14 allowed access to all life's metabolism, down to the position of each carbon atom in a biomolecule—in a protein, a sugar, a fat, or a nucleic acid. Such detail in metabolic cycles has brought about advances in nutrition, understanding of disease, and administration of drugs.

Studies of photosynthesis led directly to the use of carbon-14 in under- standing photosynthesis in the ocean. Melvin Calvin and his team at Berkeley helped fertilize the mind of Einer Steemann Nielsen and his associates in Denmark. Steemann Nielsen added carbon-14 to ocean- water samples from around the world and, in 1952, set off the widespread use of the "carbon-14 method" to understand the magnitude and variability of photosynthetic production in the ocean. Calvin and his group used microalgae as their experimental model, and microscopic autotrophs dominate ocean ecosystems. On land, photosynthesis outcomes (plant growth) can be observed directly, but that is not possible in the ocean (or aquatic) realm—hence the value of carbon-14.

The other thread makes use of the natural abundance of carbon-14. Like other scientists of the time, Willard Libby redirected his research during World War II. Soon after the war, however, he solved the problems of measuring the minute quantities of carbon-14 produced in the upper atmosphere as a result of cosmic rays, and thereby caused a revolution in anthropology, archeology, geology, oceanography, and climate science. Anything with carbon in it could now be given a sort of "death certificate."

Studies of the natural abundance of carbon-14 proceeded, in the 1950s through the 1970s, to work out the circulation times in the deep ocean and the ocean's biological and chemical cycling of carbon. Dating of archeological artifacts continued apace. In the 1980s, the development of accelerator mass spectrometry (AMS) engendered yet another revolution, this time in the assessment of carbon-14. Instead of waiting for carbon-14 to produce a scintilla of light or a "tick," signifying a decay electron, AMS measured the amounts of carbon-14 directly. AMS also enabled the study of proxies of past climates—foram shells, cave deposits, lake sediments, and the like.

Assessing and figuring out the dynamics of the natural abundance of carbon-14, however, has not been without complications. First, it was discovered that it could not be assumed that the production rate of carbon-14 by cosmic rays was constant. Artifacts or samples might appear to be younger or older than they actually were, or else dates could only be assigned over a range of years. Conversely, the sometimes variable cosmic production of carbon-14 could be turned around and used as an indicator of the drivers of climate change.

At first sight, we might think that experiments in which carbon-14 is added as a tracer would be more easily amenable to interpretation. That is true to large extent for research conducted in the lab. However, as with studies of the natural abundance of carbon-14, once we take our measurements out into the environment—and, in particular, the ocean—things become difficult and contentious. The carbon-14 method proved powerful in the beginning, but with more understanding and better measurements of ocean ecosystems, fractures in the outcomes from the carbon-14 method began to appear, and these were only "resolved" 30–40 years after the method was first introduced. Even at this writing there remain significant issues, questions, and puzzles regarding the productivity of the ocean. No doubt the vastness of the ocean realm and its "sampling problem" inhibits progress, but also feeds the creative spirit in oceanographers.

These puzzles mirror those surrounding carbon-14 as it occurs in nature. Because it is so fundamental, the natural abundance of carbon-14 in the environment is subject to all the variability and dynamics of both the physical and biological worlds. For example, the carbon-14 content in a fish will register what it has been eating, which in turn will reflect the chemistry of the surrounding water, which will be influenced by how the ocean has mixed. All these effects will be subservient to solar, Earth magnetism, and galactic variability that control the amount of carbon-14 produced in the atmosphere before it enters the ocean. The natural world is an unforgiving arena for science.

Having said all that, we have gained substantial understanding about the natural world over the past 60–70 years, in no small part because of carbon-14. Carbon-14 has created revolutions in science. That science has matured such that we expect advances more incrementally than abruptly (but who knows?). The future with carbon-14 portends as much stimulating science as the previous half-century.

Some of the more intriguing developing applications for carbon-14 have to do with extending the capabilities of AMS. Newer systems can operate at lower power, with a single stage of acceleration and reduced size. Instead of a basement-filling instrument (such as the Keck facility at UCI), these single-stage systems are laboratory size. As alluded to in chapter 8, there is an emerging area of biomedicine referred to as "bomb-pulse," utilizing the rapid change in atmospheric carbon-14 concentrations that resulted from nuclear tests (before the Limited Nuclear Test Ban Treaty of 1963). How humans replace or renew neurons or heart muscle can now be quantified. Bomb-pulse methods have also been valuable to forensic medicine.

These new mini-revolutions have happened within the last 10 years or so. Expect many more in the coming years. So, stay tuned.

APPENDIX 1

LIST OF NOBEL PRIZES MENTIONED IN THE TEXT

YEAR	NOBEL	NAME(S)	DESCRIPTION
1903	Physics	Antoine Henri Bequerel	Discovery of radioactivity
		Pierre and Marie Curie	Research on radioactive phenomena
1908	Chemistry	Ernest Rutherford	Chemistry of radioactive substances
1911	Chemistry	Marie Curie	Discovery of radium and polonium
1931	Medicine	Otto Warburg	Discovery and mode of action of the respiratory enzyme
1934	Chemistry	Harold Urey	Discovery of heavy hydrogen
1939	Physics	Ernest O. Lawrence	Invention and development of the cyclotron
1943	Chemistry	George de Hevesy	Isotopic tracers in chemical processes
1951	Physics	Edwin M. McMillan and Glenn T. Seaborg	Chemistry of transuranic elements
1953	Medicine	Hans A. Krebs	Discovery of the citric acid cycle
1959	Physics	Emilio G. Segré and Owen Chamberlain	Discovery of the antiproton
1960	Chemistry	Willard F. Libby	Carbon-14 dating
1961	Chemistry	Melvin Calvin	Carbon assimilation in plants

PERIODIC TABLE OF THE ELEMENTS

1	2	3	4	5	6	7	8	9	10	11	12	13	14	15	16	17	18
1 H 1.008																	2 He 4.0026
3 Li 6.94	4 Be 9.0122											5 B 10.81	6 C 12.011	7 N 14.007	8 O 15.999	9 F 18.998	10 Ne 20.180
11 Na 22.990	12 Mg 24.305											13 Al 26.982	14 Si 28.085	15 P 30.974	16 S 32.06	17 Cl 35.45	18 Ar 39.948
19 K 39.098	20 Ca 40.078	21 Sc 44.956	22 Ti 47.867	23 V 50.942	24 Cr 51.996	25 Mn 54.938	26 Fe 55.845	27 Co 58.933	28 Ni 58.693	29 Cu 63.546	30 Zn 65.38	31 Ga 69.723	32 Ge 72.630	33 As 74.922	34 Se 78.97	35 Br 79.904	36 Kr 83.798
37 Rb 85.468	38 Sr 87.62	39 Y 88.906	40 Zr 91.224	41 Nb 92.906	42 Mo 95.95	43 Tc (98)	44 Ru 101.07	45 Rh 102.91	46 Pd 106.42	47 Ag 107.87	48 Cd 112.41	49 In 114.82	50 Sn 118.71	51 Sb 121.76	52 Te 127.60	53 I 126.90	54 Xe 131.29
55 Cs 132.91	56 Ba 137.33	57-71 *	72 Hf 178.49	73 Ta 180.95	74 W 183.84	75 Re 186.21	76 Os 190.23	77 Ir 192.22	78 Pt 195.08	79 Au 196.97	80 Hg 200.59	81 Tl 204.38	82 Pb 207.2	83 Bi 208.98	84 Po (209)	85 At (210)	86 Rn (222)
87 Fr (223)	88 Ra (226)	89-103 #	104 Rf (265)	105 Db (268)	106 Sg (271)	107 Bh (270)	108 Hs (277)	109 Mt (276)	110 Ds (281)	111 Rg (280)	112 Cn (285)	113 Nh (286)	114 Fl (289)	115 Mc (289)	116 Lv (293)	117 Ts (294)	118 Og (294)

* Lanthanide series

57 La 138.91	58 Ce 140.12	59 Pr 140.91	60 Nd 144.24	61 Pm (145)	62 Sm 150.36	63 Eu 151.96	64 Gd 157.25	65 Tb 158.93	66 Dy 162.50	67 Ho 164.93	68 Er 167.26	69 Tm 168.93	70 Yb 173.05	71 Lu 174.97

 # Actinide series

89 Ac (227)	90 Th 232.04	91 Pa 231.04	92 U 238.03	93 Np (237)	94 Pu (244)	95 Am (243)	96 Cm (247)	97 Bk (247)	98 Cf (251)	99 Es (252)	100 Fm (257)	101 Md (258)	102 No (259)	103 Lr (262)

NOTES

PREFACE

1. The speech quoted here uses the correct nomenclature for referring to an isotope of an element. However, for readability, in this book I will use "carbon-14" instead of the conventional and more scientific "^{14}C."

1. DISCOVERY

1. The dates reported for the discovery of carbon-14 vary. Kamen (1963) says, "On 15 February, during a particularly violent storm . . ." Kamen (1972) states, "As the operation drew to an end in the early morning hours of February 15 . . ." However, in his 1986 book, *Radiant Science, Dark Politics*, Kamen gives the date as February 19, a little closer to the timing of the Nobel presentation described later in this chapter. The Berkeley newspaper from each of those dates, however, gives no indication of any storm. There is a news report of a severe storm with flooding in low-lying areas the night of February 27. This last date is probably correct, since it is immediately prior to the announcement at the Nobel presentation on February 29.
2. Technetium only occurs on Earth as the result of the natural decay of uranium and thorium isotopes. It has a short enough half-life (4.2 million years) compared to the birth of the sun and solar system, approximately 4.5 billion years ago (primordial production), that it would have decayed completely away.
3. The National Science Foundation (NSF) was created to support "basic," or "curiosity-driven," research, as detailed in a report by Vannevar Bush (1945) called *Science, the Endless Frontier*, commissioned by President Franklin Delano Roosevelt. As long as the scientific questions that researchers asked were considered important by their peers, a project could be funded. The idea behind the NSF was that the U.S. government would have a group of scientists to call on in case of national emergency, a lesson learned during World War II. Furthermore, "creation of new scientific knowledge" was seen as

essential to support industrial applications. The Department of Energy, on the other hand, is a mission-oriented agency that funds "outcomes-driven" research. Other mission agencies include the Office of Naval Research and the National Aeronautics and Space Administration (NASA).

4. See Kamen 1986, 133.

2. DISCOVERY'S WAKE

1. The oceanography experiments, "IronEx-1" and "IronEx-2," were designed to test the hypothesis, proposed by John Martin, that the availability of iron in seawater limited productivity in the Equatorial Pacific Ocean (see Coale et al. 1996). In these experiments, iron sulfate was spread over the surface of the ocean and the biological response measured over the next several days.

2. Accounts of Samuel Ruben's accidental death vary. Benson (2002) has Ruben carrying the apparatus outside the lab, for safety reasons, and succumbing on the lawn. The account here is from Kamen's 1986 book, *Radiant Science, Dark Politics*.

3. Thermodynamically, dissolved inorganic carbon can only exist as carbon dioxide at pH values less than 4, equivalent to the acidity of vinegar. To remove all inorganic carbon (carbonate and bicarbonate ions, for example) from a water sample, it therefore needs to be acidified.

3. THE "INVISIBLE PHENOMENON"

1. Exemplifying the switch to digital, Kodachrome, the favored film of professionals, was discontinued by Kodak in early 2017, as was the process used to develop the film. Who knows how many undeveloped rolls of Kodachrome are still out there?

2. X-rays occur whenever high-energy particles (electrons or ions) strike a material. As such, they are a sort of secondary radiation.

3. A photon can be described as a particle, or package, of light; it is sometimes called a quantum. Packages of light, plural, then become quanta.

4. People who lived in the 1950s and early 1960s in areas between the Cascade and Rocky Mountains (parts of Oregon, Washington, Nevada, and Utah) and were exposed to nuclear testing (above or below ground) or accidents at nuclear facilities are known as "Downwinders." The term also refers to those who were employed in nuclear facilities or uranium mines—that is, anyone who could have been exposed to increased levels of radiation, irrespective of accidents, and therefore would have been at an increased risk for adverse health effects. The Radiation Exposure Compensation Act of 1990 indemnified those exposed with compensation of $50,000–100,000, depending on the type and duration of exposure. January 12, 2012, was the first "National Downwinders Day."

4. DATING

1. The common term for carbon-14 dating is "radiocarbon dating"; however, I will stay with the terminology I have used throughout.
2. Libby (1955) describes his laboratory's methodology. Libby (1970) also presents a brief history and discusses the problems as perceived in the late 1940s and early 1950s.
3. A more accurate half-life for carbon-14 was determined to be 5,730 years (Godwin 1962). However, by international agreement, Libby's half-life is what is reported by carbon-14 laboratories for calibration to astronomical years.
4. I thank Ellen Druffel—introduced later, in chapter 7—for telling me about Hans Suess's three "revolutions."
5. There are two other assumptions, as well, with carbon-14 dating (see Hedman 2008). One is that any sample has been uncontaminated with other organic matter since its death. This assumption is easily accepted as long as care is taken with sample collection and preparation. The other assumption is that the carbon-14 in the organic material is broadly representative, geographically and across species. This assumption is also easily justified.
6. E. H. Willis (1996) calls Hessel de Vries the "unsung hero of [carbon-14] dating" for his methodological advances and for his ideas regarding the variations in carbon-14 dates. De Vries wrote a short book, *Variation in Concentration of Radiocarbon with Time and Location on Earth*, published by Akademie Van Wet in 1958. Some believe he might have shared the 1960 Nobel Prize given to Willard Libby. Sadly, however, the intensity he exhibited in his science carried over to his personal life. He fell in love with a research assistant, and when his overtures, including abandoning his family, went unanswered, he murdered the woman and then committed suicide, in 1959.
7. I have skipped a step here, for brevity. Cosmic rays first interact with atmospheric nuclei to generate neutrons, which then enter the nitrogen-14 nucleus, which emits a proton to make carbon-14.

5. PHOTOSYNTHESIS

1. Some geologists further divide the Hadean, calling the period before the proto-Earth collision that formed the Earth-moon system the Chaotian, after the Greek mythological first being, Chaos.
2. Sixty or so years ago, Stanley Miller, a student at the University of Chicago, created a laboratory system that perhaps approximated the early atmosphere and ocean chemistry on Earth. His "atmosphere" contained ammonia, water vapor, and methane. His "ocean" also contained sulfur hydroxide. He used a spark to simulate lightning. After a few days, organic molecules appeared—amino acids, acetic acid, and others, some of the basic building blocks of life. Miller demonstrated that organic, or biotic, molecules could be

created from abiotic forces. How these molecules assembled themselves into a replicating, information-carrying entity—that is, a living cell—remains elusive.

3. Bacteria and archaea are similar in form and function but occupy two separate domains of life. Bacteria are the template for the evolution of eukaryotic cells, the Eukaryota, cells with intracellular organelles, such as nuclei, mitochondria, and chloroplasts. Archaea have diversified and evolved in different directions, but are no less distributed in today's environments than are the bacteria.

4. It is said that George Washington's use of variolation contributed to the success of his armies during the latter parts of the American Revolutionary War.

5. The title of Ingenhousz's seminal work pretty much says it all: *Experiments Upon Vegetables: Discovering Their Great Power of Purifying the Common Air in Sunshine, and of Injuring It in the Shade and at Night to which is joined a new method of examining the Degree of Salubrity of the Atmosphere.*

6. CALVIN'S CYCLE

1. Like all of biological classification, this arrangement remains unsettled, provisional, and subject to change.

2. Martin and Synge won the Nobel Prize in Chemistry in 1941 for the development of two-dimensional paper chromatography.

3. Benson and Bassham seem to have taken philosophically that Calvin never shared his Nobel Prize with them.

7. SCINTILLATIONS AND ACCELERATIONS

1. Although I distinctly remember that these two isotopes, americium-241 and cesium-137, were used in the Dalhousie Department of Oceanography's LSC in the 1970s, I have been unable to verify that this was the case. LSC models and manufacturers change frequently, and institutional memories are not long. Cesium-137 is used for standardizing counts of tritium (3H). Americium-241 has been used at least recently to assess quenching of carbon-14 (Tudyka et al. 2017).

2. Spinthariscopes can be found for sale at several online outlets. Although I didn't appreciate it at the time, when I was eight years old, I sent away (15 cents + a cereal box top) for a "Lone Ranger Atom Bomb Ring," which was a small spinthariscope with a trace of polonium within the "silver bullet." It also had a secret message compartment. How the Lone Ranger was connected to nuclear chemistry was otherwise unclear. I remember being underwhelmed, even at that young age.

3. The actual equation for standardization is [cpm(std+sample) − cpm(sample)]/dpm(std).

4. Attributed to "Unknown" by Byrne (1986).

5. The other two components of the triad are the land-based ICBMs and the fleet of B-52 bombers.

6. "The 1977 Archaeometry Conference was held in Philadelphia March 16–19, and I have only two memories of it. One is of the fabulous dinner held in the Egyptian Room . . . The second is of a quiet conversation with Gordon Brown, who took me aside to tell me that a colleague, Roy Middleton, had just shown that negative nitrogen atoms were too unstable to accelerate in a tandem [accelerator]. On the plane ride home, the tandem/Enge plan became truly pertinent. With its external negative ion source and the possibility of simultaneously accelerating more than one isotope, the tandem could be configured as a large isotope ratio mass spectrometer . . . If nitrogen couldn't get out of the negative source, a major obstacle was removed" (Nelson 2010). I thank John Southon for finding this reference.

7. Other U.S. facilities include the Center for Accelerator Mass Spectrometry (CAMS) at Lawrence Livermore National Laboratory, the National Ocean Sciences Accelerator Mass Spectrometry (NOSAMS) at the Woods Hole Oceanographic Institution, and the Arizona Accelerator Mass Spectrometry (AMS) Laboratory at the University of Arizona.

8. THE SHROUD OF TURIN AND OTHER RELICS

1. Turkish archeologists now believe that the Barians may have made off with the wrong relics (Ensor 2017).

2. The standard deviation is a statistical term that quantifies how closely samples from a population cluster around a mean value. A small value of the standard deviation is saying, essentially, that, here, the measured dates agree.

3. Gove also mentions that some of the analysts were hoping for a first-century date for the shroud.

4. Some South American mammalian fossils from groups similar to those found in North America are dated earlier; instead of northern mammals invading South America, some scientists there argue the opposite, that South American fauna may have invaded the north.

5. Thomas (2000) is very good on the controversies surrounding Kennewick Man.

6. An internet search will identify many pages and videos of the analyses of Ötzi's remains. Detailed forensic analysis suggests that he had an arrow point in his back and a head wound, and also blood on his arrows from individuals other than himself. His final hours were no doubt violent.

7. A nice summary of recent applications of bomb-pulse methods is at http://www.pbs.org /wgbh/nova/next/body/bomb-pulse/.

9. OCEAN CIRCULATION

1. In oceanographic parlance, a "station" is anywhere a research vessel stops to make measurements: a hydrographic profile, a net tow, a bottom grab, a core, etc.

2. This is an old Breton prayer. Back in his day, Admiral Hyman Rickover gave plaques engraved with the prayer to new submarine captains. A poem by Winfred Ernest Garrison has as a first stanza, "Thy sea, O God, so great/My boat so small/It cannot be that any happy fate Will me befall/Save as Thy goodness opens paths for me/Through the consuming vastness of the sea."

3. The samplers are triggered electronically from the ship's lab to enclose a water sample at the depth of the CTD rosette system.

4. Quoted from *Philosophical Transactions of the Royal Society* 47: 214 by Kennedy (1929).

5. For a good summary of Thompson's life and myriad contributions, see *Popular Science Monthly* 9 (June 1876) at https://en.wikisource.org/wiki/Popular_Science_Monthly /Volume_9/June_1876/Sketch_of_Benjamin_Thompson_(Count_Rumford). Also Brown (1968) edited Thompson's complete works in five volumes.

6. "Black and Tan" were the colors, black trousers and khaki shirts, of the Royal Irish Constabulary Reserve Force, a collection of former British soldiers from World War I sent to Ireland to put down, often horribly, the movement for Irish independence. In Ireland, mixing two beers in a glass is instead called a "half and half."

7. Haline comes from the halogen element column in the periodic table, which includes highly reactive elements like chlorine, bromine, and fluorine that, when combined with metals, become salts—for example, sodium chloride, or table salt.

8. Martin made this observation during a committee meeting in the late 1980s for planning the Joint Global Ocean Flux Study, an international ocean science program, which was conducted later, in the 1990s. I'm sure it was not an isolated comment.

10. CARBON-14 IN THE OCEAN

1. That situation has of course changed. There are now a multitude of companies offering technological solutions to oceanographic sampling and analysis.

2. The problem of obtaining accurate temperatures at particular and accurate depths was solved early in the twentieth century. A set of two precision thermometers attached to a water sampler would flip, or reverse their direction, when a "messenger" (a lead weight that traveled along the wire) struck an actuator. (The messenger would also close valves to trap a water sample.) The glass tube containing the mercury in the thermometers has a break in it such that the reversing action preserves the level of mercury indicating the temperature when it is flipped. Of the two thermometers, one is exposed to water pressure, and the other is protected from that pressure. From the difference between the two thermometer readings, protected and unprotected, the water pressure can be calculated. And since pressure is proportional to water depth, an accurate depth of the sampler can be determined.

3. Radioactive decay being a random process, as discussed in chapter 4, it is impossible to predict the decay of an individual carbon-14 atom. Averaging over many, many decay events produces the decay constant of 8,300 years.

4. Even though the *Vema* had "New York" on its stern transom, identifying its home port, it was more efficient to send crews and scientific parties to wherever it was in the world rather than have it return home. Consequently, many Lamont scientists never saw the *Vema* unless they sailed on it.

5. See note 2, above, about the accurate determination of sampler depth.

6. IDOE (1971–1980) was the third "big science" oceanography program funded by NSF. It exceeded by an order of magnitude, in funding, the previous International Indian Ocean Expedition (IIOE, 1959–1965) and expanded the scope of ocean investigations over that of the International Geophysical Year (IGY, 1957).

7. A brief history of GEOSECS is in Broecker's memoirs (2012). The steering and executive committees of GEOSECS read like a "who's who" of late-twentieth-century chemical oceanography.

8. The National Science Foundation produced an hour-long video about GEOSECS that, at this writing, is available on YouTube ("Rivers of the Sea: Global Ocean Survey Studies"). For those of us in a later generation of oceanographers, the video offered an opportunity to see some of the prominent names in ocean geochemistry—among them, Broecker, Peter Brewer, Pierre Biscaye, and Ray Weiss—in their younger, more hirsute years, and parading 1970s clothing styles. GEOSECS cruises were long and, I imagine, arduous. The measurements and procedures had to be completely consistent, location to location, day after day, with the "big picture," the overall goal, continually kept in mind. That said, and maybe for a bit of relief, the video recounted an equator-crossing ceremony, a long-standing seagoing tradition, that "anointed" that cruise's chief scientist, Wally Broecker.

9. Another sampler was also used during GEOSECS, appropriately called the "keg sampler." This was a beer keg, holding 60 liters, with mechanisms for flushing and closure similar to the Gerard barrel; see Young et al. (1969).

10. GEOSECS was fairly successful, although it showed limitations as a first global sampling program. It was followed in the 1980s by Transient Tracers in the Ocean (TTO), which looked at the isotopes produced by nuclear testing in terms of ocean mixing processes, focusing on the Atlantic Ocean. In the 1990s, the World Ocean Circulation Experiment (WOCE) carried on the sampling of the ocean in much the same manner; its objective was to determine ocean circulation from the distribution of geochemical properties instead of from, for example, actual measurements of ocean currents. In this century, the mantle of global sampling was taken up by Climate Variability and Predictability (CLIVAR), also an international program.

11. Cherrier et al. (1999) analyze bacterial nucleic acids for their $\Delta^{14}C$ signatures and find that bacteria metabolize perhaps ~17% of the old DOC.

11. OCEAN FERTILITY

1. ASLO recently shed the "American" part of the society's name, given that now many members come from Asia and Europe. It is now the Association for the Sciences of Limnology and Oceanography, chosen in order to keep the acronym ASLO.

2. Hank Stommel studied theology at the Yale Divinity School before he became an oceanographer. He wrote a well-cited paper with the title "Varieties of Oceanographic Experience" (1963), echoing the famous William James title *Varieties of Religious Experience*. Fritz Fuglister, a colleague of Stommel's at Woods Hole Oceanographic Institution, was an artist. Neither Stommel nor Fuglister had a Ph.D.; together with Val Worthington, also an oceanographer at Woods Hole, they formed the Society of Subprofessional Oceanographers, or SOSO.

3. As a later example of this approach, John H. Ryther, working at Woods Hole Oceanographic Institute, wrote the classic "Photosynthesis and Fish Production in the Sea" (1969).

4. Levels of dissolved oxygen in water indicate how unbalanced the chemistry of the water might be. Too much organic matter in the water with too little exposure to atmospheric exchange means that bacteria will metabolize the organic matter, consuming any available oxygen in the process.

5. Gordon Riley was my Ph.D. adviser at Dalhousie University, Halifax, Nova Scotia, in the 1970s.

6. The *Galathea* Expedition has a Facebook page: https://www.facebook.com/Galathea2/.

7. Steemann Nielsen (1952) is usually cited, but the correct citation for the method is Steemann Nielsen (1951).

8. The actual equation is $CA = [DIC] \times (DPMf/DPMa)/SampleVolume$. CA is the carbon assimilated over the time of the incubation, DPMf is the radioactivity of the filter, DPMa is the radioactivity added to the sample, and DIC is the quantity of dissolved inorganic carbon.

9. This was perhaps a first attempt at such a calculation, and an ambitious undertaking given the limited knowledge of the oceans, the terrestrial biome, lakes, wetlands, etc., at the time. Riley, in later years, distanced himself from the estimates, but the paper was revived, so to speak, by his Ph.D. adviser, G. Evelyn Hutchinson, when a collection of Riley's papers was compiled into a book (Wroblewski 1982), for which Hutchinson wrote the introduction.

10. Steemann Nielsen's initial calibration of his isotope was off by a factor of 1.45, and his early values could be adjusted by this amount (Sondergaard 2002). It is never clear whether the correction has been applied in published values, hence the range given here.

11. Ryther (1956), in particular, discusses the issues, both on his own and in the experiments completed with his colleague Ralph Vaccaro (Ryther and Vaccaro 1954).

12. These stations are the Bermuda Atlantic Time-series Study (BATS) and the Hawaii Ocean Time-series (HOT), both of which have recorded data since 1988 and continue to do so today.

13. Ernst Haeckel collected individual planktonic organisms from rowboats and carefully described each animal, producing extraordinarily detailed illustrations. Victor Hensen towed plankton nets from oceangoing research vessels and published mostly numbers of organisms. Hensen was quantitative while Haeckel was qualitative. Hensen said

organisms were distributed evenly, Haeckel the opposite. They engaged in a virulent, at times personal, debate about how the ocean's plankton should be studied, and their differences resonate today. In the end, each was right about some things, wrong about others (see Kunzig 2000). Today, biological oceanographers follow more Hensen's approach than Haeckel's, following the legacy of a trophic model and grouping the various components (autotrophs, heterotrophs, etc.). However, it might also be said that today's emphasis on microbial genes, with its more qualitative analyses (what species are there?) follow Haeckel's approach.

14. My only connection, if you can call it that, to E. Steemann Nielsen is that the first publication of my Ph.D. research appeared in the same issue, and just below, Steemann Nielsen's last paper, in *Marine Biology*, volume 46 (1978). I still have my personal copy of that issue.

15. These were the famous "big bag" experiments, the first one of which was done in the early 1960s at Nanaimo, on Vancouver Island, British Columbia. See McAllister et al. (1961).

16. Fluorescence microscopy revolutionized counts of microbial cells in seawater samples. Bacterial cells were labeled with a dye; the dye entered the cells and combined with the cell's DNA. The microscope illuminated the sample (on a slide) with a wavelength of light that would excite fluorescence of the dye, each treated cell showing up as a colored dot against the background, making for easier enumeration. Phytoplankton cells in the sample, by virtue of their chlorophyll-a pigment, fluoresced naturally.

12. RESOLUTION: PLANKTON RATE PROCESSES IN OLIGOTROPHIC OCEANS

1. I found out later that the procedure used at the arrival of the *Melville* after the PRPOOS cruise was the protocol from "Operation SWAB," a program run for the oceanographic fleet of research vessels by the Rosenstiel School of Marine and Atmospheric Sciences at the University of Miami. The same protocols are used today, or at least were up to 2010 (the most recently available version).

2. I mentioned these NSF-sponsored JGOFS programs in chapter 11. All aspects of the HOT program are described at http://hahana.soest.hawaii.edu/hot/. The Bermuda Atlantic Time-series Study (BATS) is the comparable program in the Atlantic and is described at http://bats.bios.edu/.

3. For example, Longhurst et al. (1995) and Marra et al. (1988) both found values 10 times those reported by the *Galathea* expedition for sites close to where the expedition sampled. Others have reported similar findings not far away (e. g., Gieskes et al. 1979).

4. Duncan Purdie is now professor of biological oceanography at the University of Southampton, England. Peter J. LeB. Williams is emeritus professor at Bangor University, Wales.

5. Michael Bender is now professor emeritus at Princeton University. Karen Grande is an internist in Rhode Island.

13. CARBON-14 AND CLIMATE

1. The physical oceanographers at Lamont used the Lamont estate swimming pool for equipment storage. They always said things like "It's down in the pool." I didn't appreciate what they meant until I had the opportunity to visit the storage room and realized I was actually in the pool itself; the walls of the storage room were the sides of the pool, and a doorway had been cut through the sidewall. A ceiling was put above the pool, and it became the floor of Lamont's cafeteria.

2. Snow accumulation is seasonal, and can be used like tree rings to age the ice cores. The clearest isotopic signals come from ice cores collected in Greenland. Oxygen isotopes, oxygen-18 and oxygen-16, are analyzed in air bubbles in the ice core and the ice (the water) itself. The oxygen isotope oxygen-18 can be compared (by ratio) to the much more abundant oxygen-16, much as carbon-14 is compared to carbon-12. The ratio of oxygen-18 to oxygen-16 gives an indicator of temperature, based on the formation of condensation droplets (H_2O) in clouds. The colder the temperature when the water vapor condenses, the less oxygen-18 is included in the water molecule, and the lower the ratio of oxygen-18 to oxygen-16.

3. Dorothy Peteet has since that time found evidence for the global occurrence of the Younger Dryas, from lakes and bogs in New York, to Atlantic Canada and Virginia, and then to the North Pacific. She edited two issues of *Quaternary Science Reviews* in the 1990s (vol. 12, no. 5; vol. 14, no. 9) devoted to the Younger Dryas.

4. Some put the origin of the phrase "three-dog night" in the southern hemisphere, attributing it to aboriginal tribes in the Australian outback.

5. Archeologists debate the origin of agriculture and its several stages. It comes down to whether agriculture developed as a response to "plenty," and therefore population pressure, or as a consequence of "scarcity" caused by abrupt climate change. Burroughs (2005, 188–192) discusses these issues. See also Bar-Yosef and Meadow (1995) and Smith (1995).

6. To summarize, the wobbles in Earth's spin and orbit are in (1) variations in the tilt off its axis (normally a little more than 23° toward or away, depending on where the Earth is in its orbit around the sun—currently 23.44°, but can vary between 22.1° and 24.5°); (2) variations in the eccentricity, or shape, of Earth's orbit; and (3) variations in precession, a change in Earth's orientation of its spin (that wobbly top). Strong validation for Milankovitch theory is presented in Hays et al. (1976).

7. As discussed in chapter 9, the formation of NADW, part of the meridional oceanic circulation, does not imply a driver.

8. The proxy used is cadmium, a metal found in foraminifera shells. Cadmium and phosphate are almost perfectly correlated, and this allows a chronology of phosphate concentrations to be constructed. In the present-day ocean, phosphate in NADW is low, but gets regenerated along the path of deep-ocean circulation, so that it becomes high in the North Pacific. Equally high values of phosphate (high cadmium) in the North Atlantic and North Pacific suggest a weakened or absent production of NADW.

9. Oceanographers use the unit, the sverdrup, where 1 Sv = 1,000,000 cubic meters per second, an easier number to grasp. Deep-water formation comes to about 20 Sv. The Sverdrup honors Harald Sverdrup, a Norwegian oceanographer who made fundamental discoveries about ocean circulation and its effect on ocean fertility.

10. With a lack of evidence for the St. Lawrence River pathway, recent data instead suggest that Lake Agassiz drained across to the Canadian Northwest to the MacKenzie River and then into the Arctic Ocean (Murton et al. 2010). The overall effect would be the same despite the changed route to the North Atlantic.

11. Although here, we are concerned with foraminifera shells indicating climate change within the time span that carbon-14 dating is possible, forams have been used as environmental indicators going back millions of years—for example, in oil exploration.

12. The depth is called the calcite compensation depth (CCD). The CCD is governed by the amount of CO_2 dissolved in the water, as well as temperature and pressure. Deeper than the CCD, calcite shells dissolve and are not preserved in sediments.

13. The other radioisotopes used to date the fossil corals are thorium-234, uranium-234, and uranium-238.

14. Stuiver et al. (1978), early on, were able to make measurements of carbon-14 back to 75,000 ybp, however they did this for only a relatively few dates, not a detailed chronology.

15. There are now records in the Pacific that also record isotopic changes spanning the Younger Dryas period. See Peteet et al. (1997).

16. Recent research (D. Peteet, personal communication) suggests that the missing carbon-14 of the Mystery Interval may be found as an increase in carbon-14 in peat deposits happening between 20,000 and 15,000 ybp.

17. We are still left with hypotheses, or guesses, about the Younger Dryas. Although draining a water body like Lake Agassiz could bring about a rapid cooling, evidence exists for a few other mechanisms, such as a dimming sun, wind shifts, and even an extraterrestrial impact (Kennett et al. 2009, Kjaer et al. 2018). A comet or meteoric impact would send dust into the atmosphere, reduce solar radiation, and melt ice all in a very short period.

REFERENCES

Barber, R. T., and A. Hilting. 2002. "History of the Study of Plankton Productivity." In *Phytoplankton Productivity: Carbon Assimilation in Marine and Freshwater Ecosystems*, ed. P. J. le B. Williams, D. N. Thomas, and C. S. Reynolds, 16–43. Oxford: Blackwell Science.

Bar-Yosef, O., and R. H. Meadow. 1995. "The Origins of Agriculture in the Near East." In *Last Hunters, First Farmers: New Perspectives on the Prehistoric Transition to Agriculture*, ed. D. T. Price and A. B. Gebauer, 39–94. Santa Fe, N.M.: School of American Research Press.

Bassham, J. A. 2003. "Mapping the Carbon Reduction Cycle: A Personal Retrospective." *Photosynthesis Research* 76: 35–52.

Bennett, C. L., R. P. Beukens, M. R. Clover, H. E. Gove, R. B. Liebert, A. E. Litherland, K. H. Purser, and W. E. Sondheim. 1977. "Radiocarbon Dating Using Electrostatic Accelerators: Negative Ions Provide the Key." *Science* 198: 508–510.

Benson, A. A. 1951. "Identification of Ribulose in $C^{14}CO_2$ Photosynthesis Products." *Journal of the American Chemical Society* 73(6): 2971–2972.

Benson, A. A. 2002. "Paving the Path." *Annual Reviews of Plant Biology* 53: 1–25.

Bergmann, O., R. D. Bhardwaj, S. Bernard, S. Zdunek, F. Barnabé-Heider, S. Walsh, J. Zupicich, K. Alkass, B. A. Buchholz, H. Druid, S. Jovinge, and J. Frisén. 2009. "Evidence for Cardiomyocyte Renewal in Humans." *Science* 324: 98–102.

Broecker, W. S., A. Mix, M. Andree, and H. Oeschger. 1984. "Radiocarbon Measurements on Coexisting Benthic and Planktic Foraminifera Shells: Potential for Reconstructing Ocean Ventilation Times Over the Past 20,000 Years." *Nuclear Instruments and Methods in Physics Research B* 5(2): 331–339.

Broecker, W. S. 1987. "The Biggest Chill." *Natural History* 97: 74–82.

Broecker, W. S. 2003. "Does the Trigger for Abrupt Climate Change Reside in the Ocean or in the Atmosphere?" *Science* 300: 1519–1522.

Broecker, W. S. 2005. *The Role of the Ocean in Climate Change.* Eldigio Press.

Broecker, W. S. 2009. "The Mysterious ^{14}C Decline." *Radiocarbon* 51: 109–119.

Broecker, W. S. 2012. "Memoirs." *Geoscience Perspectives* 1(2).

Broecker, W. S., R. Gerard, M. Ewing, and B. C. Heezen. 1960. "Natural Radiocarbon in the Atlantic Ocean." *Journal of Geophysical Research* 65: 2903–2931.

Broecker, W. S., D. Peteet, and D. Rind. 1985. "Does the Ocean-Atmosphere System Have More Than One Stable Mode of Operation?" *Nature* 315: 21–26.

Brown, S.C. (Ed). 1968. *Collected Works of Count Rumford. Volume 1. The Nature of Heat.* Cambridge, Mass.: Belknap Press of Harvard University Press.

Buchanan, B. B., and J. H. Wong. 2013. "A Conversation with Andrew Benson: Reflections on the Discovery of the Calvin–Benson Cycle." *Photosynthesis Research* 114: 207–214 10.1007/s11120-012-9790-1.

Burroughs, W. J. 2005. *Climate Change in Prehistory: The End of the Reign of Chaos.* Cambridge: Cambridge University Press.

Bush, V. 1945. *Science, the Endless Frontier: A Report to the President.* Washington, D.C.: U.S. Government Printing Office.

Byrne, R. 1986. *The Third and Possibly the Best 637 Things Anybody Ever Said.* New York: Atheneum.

Calvin, M. 1989. "Forty Years of Photosynthesis and Related Activities." *Photosynthesis Research* 21: 3–16.

Calvin, M. 1992. *Following the Trail of Light: A Scientific Odyssey.* Washington, D.C.: American Chemical Society.

Cherrier, J., J. E. Bauer, E. R. M. Druffel, R. B. Coffin, and J. P. Chanton. 1999. "Radiocarbon in Marine Bacteria: Evidence for the Ages of Assimilated Carbon." *Limnology and Oceanography* 44: 730–736.

Committee on Abrupt Climate Change. 2002. *Abrupt Climate Change: Inevitable Surprises.* Washington, DC: National Academy Press.

Coale, K. H., K. S. Johnson, S. E. Fitzwater, R. M. Gordon, S. Tanner, F. P. Chavez, L. Ferioli, C. Sakamoto, P. Rogers, F. Millero, P. Steinberg, P. Nightingale, D. Cooper, W. P. Cochlan, M. R. Landry, J. Constantinou, G. Rollwagen, A. Trasvina, and R. Kudela. 1996. "A Massive Phytoplankton Bloom Induced by an Ecosystem-scale Iron Fertilization Experiment in the Equatorial Pacific Ocean." *Nature* 383: 495–501.

Damon, P. E., D. J. Donahue, B. H. Gore, A. L. Hatheway, A. J. T. Jull, T. W. Linick, P. J. Sercel, L. J. Toolin, C. R. Bronk, E. T. Hall, R. E. M. Hedges, R. Housley, I. A. Law, C. Perry, G. Bonani, S. Trumbore, W. Woelfli, J. C. Ambers, S. G. E. Bowman, M. N. Leese, and M. S. Tite. 1989. "Radiocarbon Dating of the Shroud of Turin." *Nature* 337(6208): 611–615.

Druffel, E. R. M. 2016. "Radiocarbon in the Ocean." In *Radiocarbon and Climate Change: Mechanisms, Applications and Laboratory Techniques,* ed. E. A. G. Schuur, E. R. M. Druffel, and S. E. Trumbore, 139–156. Basel, Switzerland: Springer International.

Eddy, J. A. 1976. "The Maunder Minimum." *Science* 192: 1189–1202.

Ensor, J. "Archaeologists in Turkey Believe They Have Discovered Santa Claus's Tomb." *Telegraph,* October 4, 2017, https://www.telegraph.co.uk/news/2017/10/04/turkish-archaeologists-believe-have-discovered-santa-clauss/.

Fairbanks, R. G., R. A. Mortlock, T.-C. Chiua, L. Cao, A. Kaplan, T. P. Guilderson, T. W. Fairbanks, A. L. Bloom, P. M. Grootes, and M.-J. Nadeau. 2005. "Radiocarbon Calibration Curve Spanning 0 to 50,000 Years BP Based on Paired ^{230}Th/ ^{234}U/ ^{238}U and ^{14}C Dates on Pristine Corals." *Quaternary Science Reviews* 24: 1781–1796.

Gieskes W. C. C., G. W. Kraay, and M. A. Baars. 1979. "Current ^{14}C Methods for Measuring Primary Production: Gross Underestimates in Oceanic Waters." *Netherlands Journal of Sea Research* 13: 58–78.

Godwin, H. 1962. "Half-Life of Radiocarbon." *Nature* 195: 984.

Gordon, A. L. 1986. "Interocean Exchange of Thermocline Water." *Journal of Geophysical Research* 91: 5037–5046.

Gove, H. E. 1999. *From Hiroshima to the Iceman: The Development and Applications of Accelerator Mass Spectrometry.* Bristol, UK: Institute of Physics.

Hain, M., D. M. Sigman, and G. H. Haug. 2014. "Distinct Roles of the Southern Ocean and North Atlantic in the Deglacial Atmospheric Radiocarbon Decline." *Earth and Planetary Science Letters* 394: 198–208.

Hays, J. D., J. Imbrie, and N. J. Shackleton. 1976. "Variations in the Earth's Orbit: Pacemaker of the Ice Ages." *Science* 194(4270): 1121–1132.

Hedman, M. 2008. *The Age of Everything: How Science Explores the Past.* Chicago: University of Chicago Press.

Hughen, K. A., J. T. Overpeck, L. C. Peterson, and R. F. Anderson. 1996. "The Nature of Varved Sedimentation in the Cariaco Basin, Venezuela, and Its Paleoclimatic Significance." *Geological Society London Special Publications* 116(1): 171–183.

Hughen, K. A., J. R. Southon, S. J. Lehman, and J. T. Overpeck. 2000. "Synchronous Radiocarbon and Climate Shifts During the Last Deglaciation." *Science* 290: 1951–1954.

Kamen, M. D. 1963. "Early History of Carbon-14." *Science* 140: 584–590.

Kamen, M. D. 1972. "The Night Carbon-14 Was Discovered." *Environment Southwest*, no. 448 (November 1972): 10–12.

Kamen, M. D. 1986. *Radiant Science, Dark Politics.* Berkeley: University of California Press.

Kennedy, A. E. C. 1929. *Stephen Hales, D. D., F. R. S.: An Eighteenth Century Biography.* Cambridge: Cambridge University Press.

Kennett, D. J., J. P. Kennett, A. West, C. Mercer, S. S. Que Hee, L. Bement, T. E. Bunch, M. Sellers, and W. S. Wolbach. 2009. "Nanodiamonds in the Younger Dryas Boundary Sediment Layer." *Science* 323: 94.

Kjær, K. H., N. K. Larsen, T. Binder, A. A. Bjørk, O. Eisen, M. A. Fahnestock, S. Funder, A. A. Garde, H. Haack, V. Helm, M. Houmark-Nielsen, K. K. Kjeldsen, S. A. Khan, H. Machguth, I. McDonald, M. Morlighem, J. Mouginot, J. D. Paden, T. E. Waight, C. Weikusat, E. Willerslev, and J. A. MacGregor. 2018. "A Large Impact Crater Beneath Hiawatha Glacier in Northwest Greenland." *Science Advances* 4(11). DOI: 10.1126/sciadv.aar8173

Kunzig, R. 2000. *Mapping the Deep: The Extraordinary Story of Ocean Science.* New York: Norton.

Libby, W. F. 1955. *Radiocarbon Dating.* 2nd ed. Chicago: University of Chicago Press.

Libby, W. F. 1961. "Radiocarbon Dating." *Science* 133: 621–629.

Libby, W. F. 1970. "Radiocarbon Dating." *Philosophical Transactions of the Royal Society London A* 269: 1–10.

Longhurst, A., S. Sathyendranath, T. Platt, and C. Caverhill. 1995. "An Estimate of Global Primary Production in the Ocean from Satellite Radiometer Data." *Journal of Plankton Research* 17: 1245–1271.

Marra, J., L. W. Haas, and K. R. Heinemann. 1988. "Time Course of C Assimilation and Microbial Food Web." *Journal of Experimental Marine Biology and Ecology* 115: 263–280.

McAllister, C. D., T. R. Parsons, K. Stephens, and J. D. H. Strickland. 1961. "Measurements of Primary Production in Coastal Sea Water Using a Large-Volume Plastic Sphere." *Limnology and Oceanography* 6: 237–258.

Morton, O. 2008. *Eating the Sun: How Plants Power the Planet.* New York: Harper Collins.

Muller, R. A. 1977. "Radioisotope Dating with a Cyclotron." *Science* 196: 489–494.

Murton, J. B., M. D. Bateman, S. R. Dallimore, J. T. Teller, and Z. Yang. 2010. "Identification of Younger Dryas Outburst Flood Path from Lake Agassiz to the Arctic Ocean." *Nature* 464: 760–763.

Nelson, D. E., R. G. Korteling, and W. R. Stott. 1977. "Carbon-14: Direct Detection at Natural Concentrations." *Science* 198: 507–508.

Nelson, E. 2010. "Personal Recollections of a Good Experiment." *Radiocarbon* 52(2–3): 219–227.

Nielson, J., R. B. Hedeholm, J. Heinemeier, P. G. Bushnell, J. S. Christiansen, J. Olsen, C. B. Ramsey, R. W. Brill, M. Simon, K. F. Steffensen, and J. F. Steffensen. 2016. "Eye Lens Radiocarbon Reveals Centuries of Longevity in the Greenland Shark (*Somniosus microcephalus*)." *Science* 353: 702–704.

Peteet, D., A. Del Genio, and K. K.-W. Lo. 1997. "Sensitivity of Northern Hemisphere Air Temperatures and Snow Expansion to North Pacific Sea Surface Temperatures in the Goddard Institute for Space Studies General Circulation Model." *Journal of Geophysical Research* 102: 23781–23791.

Rabinowitch, E. I. 1948. "Photosynthesis." *Scientific American* 179: 24–35.

Renfrew, C. 1973. *Before Civilization: The Radiocarbon Revolution and Prehistoric Europe.* New York: Knopf.

Riley, G. A. 1944. "The Carbon Metabolism and Photosynthetic Efficiency of the Earth as a Whole." *American Scientist* 32: 129–134.

Riley, G. A. 1953. "Letter to the Editor." *Journal du Conseil International pour l'Exploration de la Mer* 19: 85–89.

Riley, G. A., H. Stommel, and D. F. Bumpus. 1949. "Quantitative Ecology of the Plankton of the Western North Atlantic." *Bulletin of the Bingham Oceanographic Collection* 12(3): 1–169.

Ryther, J. H. 1956. "The Measurement of Primary Production." *Limnology and Oceanography* 1: 61–70.

Ryther, J. H. 1969. "Photosynthesis and Fish Production in the Sea." *Science* 166: 72–76.

Ryther, J. H., and R. F. Vaccaro. 1954. "A Comparison of the Oxygen and ^{14}C Methods of Measuring Marine Photosynthesis." *ICES Journal of Marine Science* 20: 25–34.

Smith, B. D. 1995. *The Emergence of Agriculture*. New York: Freeman.

Sondergaard, M. 2002. "A Biography of Einer Steemann Nielsen: The Man and His Science." In *Phytoplankton Productivity: Carbon Assimilation in Marine and Freshwater Ecosystems*, ed. P. J. le B. Williams, D. N. Thomas, and C. S. Reynolds, 1–15. Oxford: Blackwell Science.

Southon, J., A. L. Noronha, H. Cheng, R. L. Edwards, and Y. Wang. 2012. "A High-Resolution Record of Atmospheric ^{14}C Based on Hulu Cave Speleothem H82." *Quaternary Science Reviews* 33: 32–41.

Steemann Nielsen, E. 1951. "Measurement of the Production of Organic Matter in the Sea by Means of Carbon-14." *Nature* 197: 684–685.

Steemann Nielsen, E. 1952. "The Use of Radio-active Carbon (C^{14}) for Measuring Organic Production in the Sea." *ICES Journal of Marine Science* 18: 117–140.

Steemann Nielsen, E., and E. Aabye Jensen. 1957. "Autotrophic Production of Organic Matter in the Sea." *Galathea Reports* 1: 49–136.

Stommel, H. 1958. "The Abyssal Circulation." *Deep-Sea Research* 5: 80–82.

Stommel, H. 1963. "Varieties of Oceanographic Experience." *Science* 139: 572–576.

Stommel, H., and A. B. Arons. 1960. "On the Abyssal Circulation of the World Ocean—I. Stationary Planetary Flow Patterns on a Sphere." *Deep-Sea Research* 6: 140–154.

Stuiver, M., C. J. Heusser, and C. Yang. 1978. "North American Glacial History Extended to 75,000 Years Ago." *Science* 200: 16–21.

Stuiver, M., and P. D. Quay. 1981. "Atmospheric ^{14}C Changes Resulting from Fossil Fuel CO_2 Release and Cosmic Ray Flux Variability." *Earth and Planetary Science Letters* 53: 349–362.

Stuiver, M., P. D. Quay, and H. G. Ostlund. 1983. "Abyssal Water Carbon-14 Distribution and the Age of the World Oceans." *Science* 219: 849–851.

Taylor, R. E. 2016. "Radiocarbon Dating: Development of a Nobel Method." In *Radiocarbon and Climate Change*, ed. E. A. G. Schuur, E. R. M. Druffel, and S. E. Trumbore, 21–44. Basel, Switzerland: Springer International.

Taylor, R. E., D. L. Kirner, J. R. Southon, and J. C. Chatters. 1998. "Radiocarbon Dates and Kennewick Man." *Science* 280: 1171.

Thomas, D. H. 2000. *Skull Wars: Kennewick Man, Archeology and the Battle for Native American Identity*. New York: Basic Books.

Thorpe, J. 1999. *The Origins of Agriculture in Europe*. London: Routledge.

Tudyka, K., S. Miłosz, A. Ustrzycka, S. Barwinek, W. Barwinek, A. Walencik-Łata, G. Adamiec, and A. Bluszcz. 2017. "A Low Level Liquid Scintillation Spectrometer with Five Counting Modules for ^{14}C, ^{222}Rn and Delayed Coincidence Measurements." *Radiation Measurements* 105: 1–6.

Vaccaro, R. F., and J. H. Ryther. 1954. "The Bactericidal Effects of Sunlight in Relation to 'Light' and 'Dark' Bottle Photosynthesis Experiments." *ICES Journal of Marine Science* 20: 18–24.

Warren, B. A., and C. Wunsch. 1981. *The Evolution of Physical Oceanography: Scientific Surveys in Honor of Henry Stommel.* Cambridge, Mass.: MIT Press.

Willis, E. H. 1996. "A Worm's Eye View of the Early Days." Cambridge Quaternary. http://www.quaternary.group.cam.ac.uk/history/radiocarbon/.

Wroblewski, J. S. (Ed.) 1982. *Selected Works of Gordon A. Riley.* Halifax, Nova Scotia: Dalhousie University.

Young, A. W., R. W. Buddemeier, and A. W. Fairhall. 1969. "A New 60-Liter Water Sampler Built from a Beer Keg." *Limnology and Oceanography* 14: 634–637.

INDEX

Page numbers in *italics* represent figures or tables.

accelerator mass spectrometry (AMS), 50, 224; AMS system, 112–15, *114*, 225; facilities, 113, 115, 235(ch7n7); Kennewick Man dated, 129; Ötzi dated, 130; Shroud of Turin dated, 119–22, *121*. *See also* carbon-14 dating

acute radiation sickness (ARS), 49

adenosine triphosphate (ATP), 94–95

AEC. *See* Atomic Energy Commission

AES, 106–7

agriculture, development of, 206, 240(n5)

Agulhas Current, 134, *135*, 142, 148

airline pilots, 46, 48, 51

air pollution, 66. *See also* fossil fuels

algae, 84. *See also* microalgae; phytoplankton

alpha particles, 10, 21, 41, 49. *See also* radiation; radioactivity

Alvarez, Luis, 108, 109–10

American Association for the Advancement of Science, 97

American Society of Limnology and Oceanography (ASLO), 167

americium-241, 102, 234(ch7n1)

AMS. *See* accelerator mass spectrometry

anaerobes, 73–74

animals, age determination with carbon-14, 130–32

Antarctic waters: Antarctic Bottom Water (AABW), 160; Antarctic Circumpolar Current, 134, *135*, 159–60; carbon-14 to carbon-12 ratio in, *156*, 156–57, *158*, 217

archaea, 71, 73–74, 234(ch5n3)

Archean Era, 71–72

archeology: carbon-14 dating of artifacts, 59, *60*, 63, 70, 122, 125–29, 224 (*see also* Shroud of Turin); dating artifacts before carbon-14, 58–59, 63, 124–25; peopling of North America, dispute over, 126–29

Arcis, Pierre d' (Bishop), 116

aromatic solvents, 103

ARS, 49

ASLO, 167, 237(n1)

Association for the Sciences of Limnology and Oceanography, 237(n1)

Atlantic Ocean: carbon-14 to carbon-12 ratio in, 155, 156–57, *158*, 159; currents and circulation, 134, *135*, 142, *144*, 144–48, 152–53, 160 (*see also specific currents*); deep-water formation,

Atlantic Ocean (*continued*)
147–48, 152–53, 160, 208–9, 217, 219
(*see also* North Atlantic Deep Water);
meridional overturning circulation
(AMOC), 143 (*see also* meridional
overturning circulation); Northwest
Atlantic, 193–95, *194*; photosynthesis
and ocean productivity studies in,
174–75, *175*, 178–79, 193–95 (*see also*
ocean fertility); salinity, 143, 149; size,
143–44; surface water age, 157–60. *See
also* ocean; oceanography
Atlantis (research vessel), 151
atmosphere: carbon-14 production in,
36, 52–53, 54–55, *55*, 224, 233(n7);
carbon-14 production variations, 62,
63, 64–65, 67–69, 224, 225; carbon
entry into ocean waters from, 152–53,
159, 161; cosmic rays and, 51–55, *55*,
224, 233(n7) (*see also* cosmic rays);
early atmosphere, 71, 233(ch5n2);
human activity's impacts on carbon-14
in, 65–67, *66*, *67*, 160–61, 225 (*see
also* nuclear tests; Suess effect);
nitrogen in (as N_2), 18, 33, 52, 111 (*see
also* nitrogen); ozone (O_3) in, 72;
photosynthesis and, 72, 93. *See also*
carbon dioxide; nitrogen; oxygen
atomic bomb. *See* nuclear tests
Atomic Energy Commission (AEC), 17,
42–44, 153
atomic particles: acceleration of, in
cyclotron, 14–16; alpha, beta, and
gamma particles, 10, 21, 41–42, 49,
232(ch3n3) (*see also specific particles*)
atoms, 6, 40. *See also* atomic particles;
isotopes; molecules, formation of;
periodic table of elements; *and specific
elements*
ATP, 94–95
automatic external standardization
(AES), 106–7

autoradiography, 86–87
autotrophs, 73, 170, *199*. *See also* bacteria;
phytoplankton

Babbage, Charles, 211
background radiation, 48–49, 56–58. *See
also* radiation
bacteria: autotrophs and heterotrophs, 73
(*see also* autotrophs; heterotrophs);
carbon-14 taken up by, in research
study, 100; chemosynthesis and
anoxygenic photosynthesis in, 73–74;
cyanobacteria, 29, 74, 75, *199*, 200;
evolution of, 72–73, 234(ch5n3);
and the microbial food web, *199*; in
the ocean, 165, 170, 175, 177, 180–81,
237(n11), 239(n16); respiration by,
177–78; Steemann Nielsen's carbon-14
method and, 175
Bainbridge, Arnold, 155
Barber, Dick, 167, 174
Bassham, James: career, 81, 99, 234(ch6n3);
photosynthesis research, 86, 90, 93, 98.
See also Bio-Organic Group
Becquerel, Henri, 7, 38–39, 48, 227
Bender, Michael, 199–200, 239(n5)
Bennett, C. L., 111
Benson, Andrew: career, 81, 98–99,
234(ch6n3); photosynthesis research,
81, 84–86, 89–90, 92, 93, 97–98 (*see
also* Bio-Organic Group); on Ruben,
12, 232(ch2n2)
Berkeley Radiation Laboratory: Bio-
Organic Group, 81, 83–93, 96–99;
carbon-14 discovered, 5–6 (*see also*
carbon-14: discovery); cyclotrons,
14–15, 16–18, 49 (*see also* cyclotron);
funding, 17–18, 20; Kamen's position
at, 10, 13, 27 (*see also* Kamen,
Martin); scientists' successes, 30–31;
synchrotron, 31; working relationships
at, 13–14. *See also* Calvin, Melvin;

Kamen, Martin; Lawrence, E. O.; Ruben, Sam

beryllium, 69, 229

beta particles, 10, 41; dangers of, 49; detection of, 44–45, *45* (*see also* Geiger counter); emitted by carbon-14, 10, 18, 42, 48 (*see also* carbon-14); scintillation counters and, 103–4. *See also* radiation; radioactivity

bicarbonate, 88, 161–62, 172–73, 175

biochemical oxygen demand (BOD), 170, 238(n4). *See also* Winkler, Ludwig Wilhelm

biomedicine: bomb-pulse research, 132, 225; radioisotopes used/needed, 17–18, 20, 27, 49

biomineralization, 163

Bio-Organic Group (photosynthesis research), 81, 83–93, 96–98, 168

biosphere, carbon-14 taken up by, 54, *55*, 61. *See also* photosynthesis

Biowatt initiative, 195

Birge, R. T., 24, 31

Black and Tan (drink), 141, 236(n6)

BOD. *See* biochemical oxygen demand

Bohr, Niels, 11

Bolling/Alleröd (climate period), 205, *206*, 210, *218*, 221

bomb-pulse research, 132, 225

bomb radiocarbon (carbon-14 from nuclear tests). *See* nuclear tests

boron (B), 8, 229

bristlecone pine tree, 68, 123, 211

Broecker, Wally, 202; carbon-14 analyses, 155; climate work, 202, 207–8, 221; and cyclotron-mass spectrometry, 115; and GEOSECS, 155, 156, 237(nn7–8); ocean conveyor belt metaphor, 148–49

bucket sea-gauge, 138–39

Bumpus, Dean, 178, 193. *See also* Riley, Gordon A.

Burney, Curt, 188–90, 192

cadmium, 240(n8)

calcite compensation depth (CCD), 213, 241(n12)

calcium carbonate, 163, 212–13, 214

Calvin, John, 116

Calvin, Melvin, *82*; Nobel Prize, vii, 28, 98, 227, 234(ch6n3); team's research on photosynthesis, vii, 28, 81, 83–93, *87*, 96–99, 168, 223, 224

Calvin-Benson-Bassham cycle (CBB cycle), 88–96, *95*, 97, 98. *See also* photosynthesis: biochemical process

carbohydrate(s), 35, 79, 83, 88, 164. *See also* sugars; *and specific molecules*

carbon (C), ix, 229; acidification and precipitation under oxidizing conditions, 22; as basis of life, ix, 35–36, 223; biomass calculation through isotope dilution, 100–101; carbon-11, 8, 19–20, 81–83, 83; carbon-12, 7, 110; carbon-13, 7, 21–22, 31, *114*, 114, 155; carbon atom, 7, *33*, 33–35; carbon molecules, *34*, 34–36; fixation through photosynthesis (*see* photosynthesis); grams per centimeter of Earth's surface, 56. *See also* carbon-14; carbon-14 dating; carbon-14 to carbon-12 ratio

carbon-14, 7, 225; AMS measurement of, 112–15 (*see also* accelerator mass spectrometry); bottom-up production from carbon-13, 21–22, 31; broad applications, ix; cyclotron-mass spectrometer measurement of, 109–12; decay rate, 54, 112, 152, 236(ch10n3); discovery, 5–6, 16, 21–24, 25–26, 31, 223, 231(n1); half-life, 23, 26, 41, 59, 221, 233(n3); identified in cloud chamber experiments, 18, 21; impact of, vii–viii, 225; importance for biochemical research, 36, 50, 96 (*see also* photosynthesis); Kamen's

carbon-14 (*continued*)
attempts to create, in cyclotron, 18–19; mean-life, 62; natural production, in atmosphere, 31, 52–53, 54–55, *55*, 62, 63, 64–65, 224, 233(n7) (*see also* atmosphere); nitrogen-14's relationship to, 7, 18–19, 31–33, 111; nitrogen contamination problem, for mass analysis, 111–12; in the ocean (*see* ocean; ocean circulation); oxidation to carbon dioxide, 36, 52–53, 54, *55* (*see also* carbon dioxide); pros and cons of long life, 107–8; radioactivity levels and safety, 7, 40, 42, 46, 48, 56, 58, 176, 183–85, 190–91; research interrupted by World War II, 26; scintillation counters and, 106, 107–8 (*see also* liquid scintillation counter); short half-life hypothesized, 18–19, 25–26, 41; top-down production from nitrogen-14 (artificial), 18, 31–33, 36; as tracer (generally), 96, 99–101, 176, 224–25 (*see also specific topics of research*); weight, vs. carbon-12, 110. *See also* carbon-14 dating; carbon-14 to carbon-12 ratio; *and specific research topics and applications*
carbon-14 dating, 123, 233(ch4n1); AMS and (*see* accelerator mass spectrometry); of archaeological artifacts, 59, *60*, 63, 70, 122, 125–29, 224 (*see also* archeology); and background radiation, 56–58; broad applications, 130–32; calibration curve, 68–69, 123; carbon-14 chronology, 211–16 (*see also* climate and climate change); carbon dioxide gas, better results with, 64; developed by Libby, vii, 54–59, *57*, *60*, 63, 70, 224, 227; human activity's effects on, 65–67, *66*, *67*; latitudinal variations tested, 58; Libby's apparatus, *57*,

57–58; nuclear tests' time stamp and, 130–32 (*see also* nuclear tests); obstacles to, 108; principles underlying, 59–62; public awareness of, viii; reservoir effect and, 65, 129; sample contamination, 233(n5); Shroud of Turin, 119–22, *121*; ultimate limit, 214; uncertainties of, 128, 129; usefulness, 70, 108; variations in carbon-14 formation rate and, 63–65, 67–69; "zero" determined, 56, 58
carbon-14 method. *See* Steemann Nielsen, Einer
carbon-14 to carbon-12 ratio: affected by changes in Earth's magnetic field, 64; altered by human activity, 65–67, *66*, *67*; AMS estimation of, 113–14; in Antarctic waters, *156*, 156–57, *158*, 217; in Atlantic waters, *155*, 156–57, *158*, 159; calibration of, 68–69; and carbon-14 dating, 63; cyclotron-mass spectrometer measurement of, 110–11; before deglaciation, 216; during periods of glaciation and deglaciation, 219–21, *221*; in seawater, *156*, 156–57, 159, 160 (*see also* ocean; ocean circulation); in sharks' eye lens nucleus, 131; in the Younger Dryas, 217, *218*
carbonate (CO_3^{-2}), 162. *See also* calcium carbonate
carbon dioxide (CO_2): anoxygenic photosynthesis and, 73; bicarbonate conversion to, 176; and the calcite compensation depth (CCD), 241(n12); Calvin team's photosynthesis research and, *87*, 87–88; dangers recognized by van Helmont, 76; and the discovery of carbon-14, 22; entry into ocean from atmosphere, 152–53, 159, 161–62; in groundwater, 214; in ice cores, as temperature indicator, 207–8; and

Kamen and Ruben's photosynthetic carbon fixation research, 19–20; molecular structure, 34, *34*, 35, 72; oxidation of carbon-14 to ($^{14}CO_2$), 36, 52–53, 54, *55*; photosynthesis role, 90–96, *95*, 97 (*see also* photosynthesis: biochemical process); reduction to carbon, 113–14; released by plants in the dark, 78. *See also* photosynthesis

carbonic acid (H_2CO_3), 161–62, 214

"Carbon Metabolism and Photosynthetic Efficiency of the Earth as a Whole, The" (Riley), 178. *See also* Riley, Gordon A.

carbon molecules (organic molecules), 33–36, *34*. *See also specific chemicals, such as* carbon dioxide

cardiomyocytes, 132

Cariaco Basin cores, 215–16, 217

caves. *See* speleothems

CBB cycle. *See* Calvin-Benson-Bassham cycle

CCD, 213, 241(n12)

cell membranes, 35

cesium-137, 102, 234(ch7n1)

cesium gun, 112, 113, 114, *114*

Challenger expedition (1870s), 140–41, 174

Charny, Geoffroi de, 116

Chatters, James, 128

Cherrier, Jennifer, 164, 237(n11)

Chlorella pyrenoidosa (green alga), 84, 88

chlorophyll, 84, 95–96. *See also* photosynthesis

chromatography. *See* paper chromatography

chronologies, creation of, 122–25. *See also* carbon-14 dating; climate and climate change

climate and climate change: Broecker, Peteet, and Rind's climate analysis, 207–8; carbon-14's role in understanding, 203, 210–21, *218*, *220*; chronologies and proxies, 202–3, 204–5, 211–16, 224 (*see also* tree rings); ice cores, climate history in, 203, *206*, 207–8, 240(n2); Milankovitch theory and, 207, 240(n6); ocean circulation and, 208–9, 217–22, *218*, *220*; ocean's carbon cycle and, 136–37; rapid/abrupt change, 205, *206*, 206–7, 209–10; today, 221–22. *See also* glaciation and deglaciation; Younger Dryas

Climate Variability and Predictability (CLIVAR), 237(n10)

cloud chamber, 9–10, 18, 21

Clovis people, 127

coccolithophores, 163

conductivity-temperature-depth profiler (CTD), 133, *137*, 137–38, 236(ch9n3)

Conqueror, The (1956 film), 42–44, *43*

copepods, 170

corals, 69, 132, 214, 241(nn13–14)

Coriolis, Gaspard-Gustave de, 145

Coriolis effect, 145–46, 160

cosmic microwave background, 123

cosmic rays (cosmic radiation), 51; and the atmosphere, 36, 51–55, 224, 233(n7) (*see also* atmosphere); average dpm, 56, 57; and carbon-14 dating, 56–58; Earth's magnetic field and, 64; human exposure to, 48, 51; latitudinal distribution, 58. *See also* background radiation

Craig, Harmon, 155

Crookes, William, 103

CTD. *See* conductivity-temperature-depth profiler

cultural ideas, multiple origins of, 125–26

curie (unit of measurement), 40

Curie, Marie, 7, 39–40, 227

Curie, Pierre, 39, 227

cyanobacteria (blue-green algae), 29, 74, 75, *199*, 200

cycloids (propulsion/steering system), 188

cyclotron, 14–17; 60-inch cyclotron
    (Berkeley), 9; carbon-14 production
    through nitrogen-14 bombardment
    in, 31–33; and the discovery of
    carbon-14, 5–6, 18, 21–23, 24 (*see also*
    carbon-14: discovery); internal vs.
    external targets, 20–21; Lawrence and
    the Berkeley cyclotron, 10–13, 17–18;
    origins, 14; radioisotope production,
    17, 20–21, 49, 82; schematic, *15*;
    synchrotron, 31
cyclotron-mass spectrometer, 109–12, 115

deglaciation, 203, 207–8, 209–10, *210*, 216,
    219–21, 241(n10). *See also* climate and
    climate change
dendrochronology, 211–12. *See also* tree
    rings
Department of Energy, 17, 232(ch1n3).
    See also Atomic Energy Commission
    (AEC)
deuterium, 20
deuterons, 5, 10–13, 19, 21–22, 23
De Vries, Hessel, 63–64, 67, 233(n6)
De Vries effect, 63–64
diatoms, 75, 162, *162*
DIC. *See* dissolved inorganic carbon
Discovery II (research vessel), 151
disintegrations per minute (dpm),
    56, 106
dissolved inorganic carbon (DIC), *156*,
    156–57, *158*, 161–62, 172, 232(ch2n3),
    238(n8)
dissolved organic carbon (DOC), 155,
    163–66, 237(n11)
DNA (deoxyribonucleic acid), 36, 49, 123,
    128–29
DOC. *See* dissolved organic carbon
dpm. *See* disintegrations per minute;
    disintegrations per minute (dpm)
Druffel, Ellen, 113, 166
drug discovery (pharmaceutical), 99

Dryas octopetala, 204–5. *See also* Younger
    Dryas

e (constant), 61
Earth: background radiation, 48–49,
    56–58 (*see also* radiation); heat
    distribution, 144–45; magnetic
    field, 58, 64, 216; origin and early
    history, 71–72; rotation and ocean
    circulation, 145–47, 160; tilt, spin, and
    orbital variations, and climate, 207,
    225, 240(n6). *See also* atmosphere;
    biosphere; climate and climate change;
    ocean
electromagnetic spectrum, *47*, 47
electrons, 10, *33*, 33–35, *34*, 41–42. *See
    also* atoms; beta particles
electron volt (eV), 16
elephants, 130–31
Ellis, Henry (Capt.), 138–39
Endeavor (research vessel), 1–3
enzymes, 92. *See also* Rubisco
Eppley, Dick, 181–82, *183*, 183–84, 186,
    190–91
Eppley, Jean, 186–87
eudiometer, *78*
Ewing, Maurice, 153
exponential relationships, 59–61, *60*, 62

Fager, Edward, 97–98
Fairbanks, Rick, 214
"Father Guido Sarducci" (character), 116
film badges, 40
film photography, 38–39, 232(ch3n1)
fish, 129, 131, *169*, 169–70, 225
flagellates, *199*
flight attendants, 46, 48, 51
fluor, 103–4
fluorescence, 103. *See also* liquid
    scintillation counter
foraminifera, 163, 208, *212*, 212–13, 217,
    240(n8), 241(n11)

fossil fuels, 65–66, *66*. *See also* hydrocarbons; petroleum
Franklin, Benjamin, 77, 147
Fuglister, Fritz, 238(n2)

Gaffron, Hans, 97–98
Galathea Expedition (1950), 174–75, *175*, 178, 195, 196, 238(n6). *See also* Steemann Nielsen, Einer
gamma radiation (gamma rays), 41, 45, *47*, 48, 49
gas (gases), 35, 76. *See also* atmosphere; carbon dioxide; nitrogen; oxygen; phosgene gas; radon
Geiger, Hans, 40–41
Geiger counter (Geiger-Müller counter), 40–41, 44–46, *45*; atmospheric radiation detected, 51, 52; and autoradiography, 86–87; Libby's carbon-14 dating apparatus, *57*, 57–59; Steemann Nielsen's carbon-14 method and, 176; Wayne's use of, 42–44
General Oceanics, 150–51
geology, 123–24
GEOSECS (Geochemical Ocean Sections Program), 155–57, 237(nn7–10)
Gerard, Robert, 153
Gerard barrel (sampler), 153–54, *154*, 155–56
glaciation and deglaciation, 203, 207–9, 214, 216, 219–21, *220*, *221*. *See also* climate and climate change
Gordon, Arnold, 133–34, 148
Gordon, R. J., vii
Gove, Harry, 122, 235(n3)
Grande, Karen, 199–200, 239(n5)
graphite, 5, 21–22, 114
Greenland, 221; ice cores, 205, *206*, 208, 217, *218*, 240(n2) (*see also* ice cores). *See also* North Atlantic Deep Water (NADW)
Greenland sharks, 131

Gulf Stream, 142, 147, 148, 152. *See also* Atlantic Ocean

Hadean Eon, 71, 233(ch5n1)
Haeckel, Ernst, 179, 238–39(n13)
Hales, Stephen, 77, 138–39
half-life (concept), 7–8, 41. *See also* specific isotopes
haline (term), 142, 236(n7)
Hawaii Ocean Time-series (HOT), 193
heart muscle cells, 132
Heinbokel, John, 186
helium (He), 34, 41, 229; helium-3 (tritium), 160
Hensen, Victor, 179, 238–39(n13)
heterotrophs, 73, 170, *199*. *See also specific groups*
Heyward, Susan, *43*, 44
Hilting, Anna, 174
HOT, 193
Hughes, Howard, 42–44
Hulu Cave (China), 215
hydrocarbons, 35, 165. *See also* fossil fuels; petroleum
hydrogen (H), *34*, 35, 73, 161–62, 229; hydrogen-3 (tritium), 106, 160. *See also* hydrocarbons; methane; water
hydrogen bomb. *See* nuclear tests

ice age. *See* glaciation and deglaciation
ice cores, 203, 205, *206*, 207–8, 217, *218*, 240(n2)
Iceman. *See* Ötzi
ice sheets, melting, 209–10, *210*, 221. *See also* glaciation and deglaciation
IDOE, 155, 237(n6)
Indian Ocean: carbon-14 to carbon-12 ratio in, 155, *156*, 156–57, *158*; expeditions, 175, 237(n6); and ocean circulation, 133–34, 148. *See also* Agulhas Current
Ingenhousz, Jan, 77–79, *78*, 234(ch5n5)

International Decade of Ocean
    Exploration (IDOE), 155, 237(n6)
ionizing radiation, *47*, 47–49. *See also*
    radiation; radioactivity
ions, 111–12
iron (Fe), 29, 73, 74, 229, 232(ch2n1)
isotopes, 6; of carbon (generally), 7;
    helium-6/lithium-6 pair, 18; ionization
    of, in mass spectrometry, 110, 111–12;
    isotope dilution, 100–101; Urey's
    work on, 20. *See also* radioactivity;
    radioisotopes; *and specific isotopes*
Iturriaga, Rodolfo, *137*
ivory, 130–31

Jeans, Sir James, ix
Joint Global Ocean Flux Study, 198

Kamen, Esther Hudson, 10, 28
Kamen, Martin: and authorship of
    carbon-14 discovery papers, 25;
    Berkeley position, 13, 27–28; and
    Berkeley's cyclotrons, 16; carbon-14
    discovered, 5–6, 16, 21–24, 25–26, 114,
    231(ch1n1); collaboration with Ruben,
    14; internal targets for cyclotron
    developed, 20–21; and Lawrence's
    platinum–deuteron bombardment
    experiment, 12–13; life, 8–10, 27–28,
    49; long-lived radioactive isotopes
    sought, 20; nitrogen-14 bombardment
    experiment, 18–19, 31–33, 52;
    photographs, *9*; photosynthetic
    carbon fixation research (with
    Ruben), 19, 20, 80, 81–83; post-
    discovery career, 26–29; radioactivity
    of, 14, 22, 28, 37; on Ruben, 29–30;
    short half-life hypothesized for
    carbon-14, 18–19, 41
Keck Carbon Cycle Accelerator Mass
    Spectrometry Laboratory, 113–15, *114*
Kennewick Man, 128–29

Knudson, Carol, 1–3
Korff, Serge A., 51–52, 54
Krebs, Hans, 96–97, 227
Krebs cycle, 96–97
krill, 170
Kristina (tech), 190

La Jolla Radiocarbon Laboratory
    (UCSD), 62–63, 65, 113
Lake Agassiz, 209, *210*, 241(n10)
lakes: carbon-14 studies, 179; sediment
    varves, 69, 124, 215; study of
    (limnology), 167; Winkler's oxygen
    measurements, 170
Lamont-Doherty Geological
    Observatory: campus, 153, 201, 202,
    240(n1); collegial relationships, 126;
    Geiger-Müller system, 45–46; ivory
    carbon-dated, 130; Muller seminar
    at, 115; oceanic carbon-14 measured,
    153–55; paleoclimate research, 201–2
    (*see also* Broecker, Wally); radiation
    safety class, 46–47; RV *Vema*, 153–54,
    237(n4); soft-money science at, ix–x.
    See also specific scientists
Landry, Michael, microbial food web
    schematic, *199*
Langdon, Chris, 1–3
Last Glacial Maximum (LGM), 203,
    208, 214, *220*. *See also* glaciation and
    deglaciation
Lavoisier, Antoine, 77
Lawrence, E. O.: and Berkeley's
    cyclotrons, 10–13, 14–15, 16–18; and
    Calvin's team, 28, 81, 96; and Kamen,
    including carbon-14 discovery, 12–13,
    20, 23–24, 25, 27; Nobel Prize, 11, 16,
    23–24, 227; photograph, *11*; research
    funding, 17–18, 20
Lenz, Emil, *140*, 140
Leonardo da Vinci, 211
Levi, Hilde, 173, 174

Lewis, Gilbert, 23

Libby, Willard, *53*; carbon-14 dating work, vii, 53–59, 63–64, 68, 70, 125, 224; Levi and, 174; Nobel Prize, vii, 227

life: beginnings of, 71–73, 233–34(ch5n2); carbon-14 taken up by, 54, 61; carbon as basis of, ix, 35–36, 72, 223; metabolic cycles understood with carbon-14 tagging, 223 (*see also* photosynthesis); as product of photosynthesis, 72 (*see also* photosynthesis). *See also specific forms of life*

light: earliest use to fuel life, 71–72, 73–74; photons, 232(ch3n3); and photosynthesis, 83, 95–96. *See also* photosynthesis; sun

lignins, 165

limestone deposits. *See* speleothems

limnology, 167. *See also* lakes

lipids (fats), 35, 223

liquid scintillation counter (LSC), 50, 102–8, *105*, 112, 115, 176, 234(ch7n1)

Litvinenko, Alexander, 49

logarithms, 8, 61

lollypop apparatus, and experiment setup, *87*, 87–88

Lorio, Phil, 46–47

Luther, Jim, *137*

macromolecules, 36

magnetic field, Earth's, 58, 64, 216

Malpighi, Marcello, 76

Manhattan Project, 20, 27, 31, 46, 54

manometry, 96–97

Maria Theresa, 77

Mark, Saint, 118

Martin, A. J. P., 85, 234(ch6n2)

Martin, John, 143–44, 236(n8)

mass spectrometry, 109, 110. *See also* accelerator mass spectrometry; cyclotron-mass spectrometer

Maunder, Edward, 64

Maunder Minimum, 64–65, 69

McCrone, Walter, 122

McMillan, Edwin (Ed), 11–13, 14, 31, 32, 227

mean-life (defined), 62

Melville, RV, 183–92, 239(n1)

meridional overturning circulation (MOC), 219, *220*, 221–22. *See also* ocean circulation

Meteor expedition (1920s), 140–41, 174

methane (CH_4), *34*, 35, 56, 58, 73

microalgae, 83–84, *87*, 87–88. See also *Chlorella pyrenoidosa*; photosynthesis

microbial food web, *199*, 200. *See also* bacteria; cyanobacteria

Middleton, Roy, 111, 235(ch7n6)

Milankovitch, Milutin, 207

Milankovitch theory, 207, 240(n6)

Miller, Stanley, 233–34(ch5n2)

MOC. *See* meridional overturning circulation

molecules, formation of, 33–35, *34*

Monte Verde, Chile, 127

Morrison, Philip, 19

Muller, Richard, 108–11, 115

Müller, Walther, 44. *See also* Geiger counter

Mystery Interval (climate period), 221, 241(n16)

NADPH, 94–95

NADW. *See* North Atlantic Deep Water

National Science Foundation (NSF), 17, 155, 183, 231–32(n3), 237(nn6, 8)

Nautilus, chambered, 131–32

Nelson, Erle, 111

neutrons, 6; and carbon-14 formation in the atmosphere, 52–53, 54, *55*; in carbon atoms and carbon molecules, *33*, 33–34, *34*; released by nuclear bombs, 66

Nicholas, Saint, 118, 235(ch8n1)
nicotinamide adenine dinucleotide phosphate (NADPH), 94–95
Niépce de Saint-Victor, Abel, 38
nitrogen (N or N$_2$), 18, 33, 52, 111, 229. *See also* nitrogen-14
nitrogen-14: abundance, 18; carbon-14's relationship to, 7, 18–19, 31–33, 52, 111; laboratory production of carbon-14 from, 18, 31–33, 36, 52, 233(n7); natural production of carbon-14 from, 31, 52, 55 (*see also* atmosphere); nitrogen contamination problem, for mass analysis, 111–12, 114, 235(ch7n6)
Nobel Prize, vii, 24, 31, 227. *See also specific scientists*
North America, peopling of, 126–29
North Atlantic Current, 147, 152. *See also* Atlantic Ocean; ocean circulation
North Atlantic Deep Water (NADW): carbon-14 inventory, 153, 160; deep-water formation, 147–48, 152–53, 208, 209, 240(n7); density and salinity, 150, 151, 152–53, 209; during glacial periods, 217, 219, 221; phosphate in, 240(n8). *See also* Atlantic Ocean; ocean circulation
North Atlantic Gyre, 146–47. *See also* Atlantic Ocean
Novello, Don, 116
NSF. *See* National Science Foundation
nuclear chemistry, 9, 33–35. *See also* accelerator mass spectrometry (AMS); cloud chamber; cyclotron; isotopes; periodic table of elements; *and specific elements and isotopes*
nuclear isomerism, 13
nuclear reactor (Oak Ridge), 27
nuclear submarines, 108–9
nuclear tests: atmospheric carbon-14 increased, 66–67, 67, 160–61, 225; bomb-pulse research, 132, 225;

oceanic carbon-14 increased, 160–61, 237(n10); radiation effects downwind, 42–44, 232(ch3n4); time stamp and carbon-14 dating, 130–32
nuclear waste disposal, 153
nucleic acid molecules, 35–36, 223

Oak Ridge National Laboratory, 27, 72
ocean: acidity, 162; age of water traced using carbon-14, 143, 152–53, 154–57, 158, 159, 165; anoxic zone, 216; atmospheric carbon's entry into, 152–53, 159, 161; bacteria in, 165, 170, 175, 177, 180–81, 237(n11), 239(n16) (*see also* bacteria); biomass, 169; bomb radiocarbon in, 160–61, 237(n10); bottom seeps, 165; carbon cycle and Earth's climate, 136–37; carbon dioxide-14 taken up by, 54; as changeable environment, 195–96; chronology for carbon-14 in, 132; and climate records, 212–14 (*see also* corals; foraminifera); coral dating using uranium isotopes, 69; deep waters, carbon-14 in, 65 (*see also* North Atlantic Deep Water); density of layers, 141–42, 151; dissolved inorganic carbon (DIC) in, 156, 156–57, 158, 161–62, 172, 232(ch2n3), 238(n8); dissolved organic carbon (DOC) in, 155, 163–66, 237(n11); distribution of carbon-14 in, 136; early ocean and the beginnings of life, 71–72, 233–34(ch5n2); *Endeavor* research (1991), 1–3; food web, 168–70, 169; food web and organic carbon, 162–64; isotope dilution as research method, 100–101; layers, 138, 141–42; majority of Earth's carbon in, 55, 55–56; microbial food web, 199, 200; mixing of layers, 159, 194 (*see also* ocean circulation); nuclear testing

and carbon-14 entry into surface waters, 67, 131–32 (*see also* nuclear tests); ocean food webs, carbon-14 in, 136, 162–63; oxygen changes in, 170–71, 177–79; particle rain rate, 180–81, 213; respiration of life forms in, 162; salinity, 133–34, 141–42, 198; sea levels, 145; sediment varves, 216; temperature, 138–39, 141–42, 198; vastness, 134–36, 174, 236(ch9n2). *See also* ocean circulation; ocean fertility; oceanography; *and specific oceans, seas, and forms of life*

ocean circulation, 133–49; carbon-14's role in understanding, 136, 149, 154–57, 224; and climate, 208–9, 217–22, *218*, *220*; conveyor belt concept, 148–49; Coriolis effect, 145–46, 160; deep-water formation, 147–48, 152–53, 160, 208–9, 217, 219, 221 (*see also* North Atlantic Deep Water); earliest hypotheses, 139, 140, *140*; identification and understanding of, 133–34, *135*, 142, 147–48, 150–52; Indonesian Throughflow, 148; meridional overturning circulation (MOC), 143, 219, *220*, 221, 222; North Atlantic circulation, 134, 142, *144*, 144–48 (*see also* Atlantic Ocean); relationship of surface and deep ocean circulations, 142–43, *144*, 152–53; temperature, salinity, and, 133–34, 141–43, 144–45, 147–49, 152–53; thermohaline circulation (TCH), 142–43; timescale, 150, 152, 154–57, *158*, 219; and water's age, 154–60, *156*, *158*; wind and, 142; winds and, 144, 145. *See also* ocean; oceanography

ocean fertility (productivity): carbon-14 method used to study, 172–78, 179–80, 197, 199, 223–25, 238(nn8, 10), 239(n15) (*see also* Steemann

Nielsen, Einer); controversy over, 176–82, 192–200, 224–25, 239(n3); current estimates, 196–97; differing approaches to studying, 195–96; ecological studies before carbon-14, 168–72, 177–78; Hawaii Ocean Time-series (HOT), 193; iron-enrichment experiments, 29, 232(ch2n1); measurement methods compared, 186; other means to estimate, 198; oxygen change method (Winkler; Riley), 170–71, 195, 197, 199–200 (*see also* Riley, Gordon A.); PRPOOS and, viii, 181–82, 185–86, 192–93, 198–200 (see also *Melville,* RV); sample contamination concerns, 176, 181, 191

oceanographic tools and methods, 150–51, 236(ch10n1); bacterial plate counts, 180; bucket sea-gauge, 138–39; carbon-14 method, 175–77 (*see also* Steemann Nielsen, Einer); conductivity-temperature-depth profiler (CTD), 133, *137*, 137–38, 236(ch9n3); Gerard barrel, 153–54, *154*, 155–56; keg sampler, 237(n9); modern data collection tools, 136; particle interceptor traps, 180–81; pipettor (auto-pipette), 191; Sverdrup units, 241(n9); Swallow floats, 150, 151–52; thermometers, 236(ch10n2); Winkler's oxygen measurement technique, 170–71, 177–79

oceanography: assaying ocean for carbon-14, 154; big bag experiments, 179–80, 188–92, 239(n15); Biowatt initiative, 195; and carbon-14 safety, 183–85, 190–91, 239(n1); core measurement protocols established, 198; differing approaches to studying, 195–96; ecological studies before carbon-14, 168–72, 177–78; GEOSECS program, 155–57, 237(nn7–10);

oceanography (*continued*)
history, 138–41, 147, 179, 238–39(n13); research vessel life, 1–3, 186, 187, 192; research voyages, 1–3, 135–36, 140–41, 151, 153–54, 174–75, 182 (see also *Melville*, RV); sampling problem, 135–36, 171–72, 195; as science, 123, 167–68, 238(n2); as soft-money science, ix–x; "station" defined, 235(ch9n1); time-series stations, 179, 238(n12). *See also* ocean; ocean circulation; ocean fertility; oceanographic tools and methods; Scripps Institution of Oceanography; Woods Hole Oceanographic Institution

oils (hydrocarbons), 35. *See also* fossil fuels; petroleum

Oppenheimer, J. Robert, 11, 19, 25–26, 28, 31

Ostlund, H. G., 156–57

Ötzi (the Iceman), 130, 236(n6)

oxidation reactions (defined), 73

oxidation-reduction, 113–14

oxygen (O), 229; anoxic waters, 215, 216; biochemical oxygen demand (BOD), 170, 238(n4); discovery, 77; in ice cores, 203, 207–8, 217, *218*, 240(n2); iron oxides, 74; ocean photosynthesis and, 170, 177–78, 181; oxygen-18, 81, 200, 207–8, 217; oxygen atom, *33*, 34 (*see also* carbon dioxide); photosynthesis and, 72, 74, 77, 80, 81, 93; Winkler's measurement method and ocean photosynthesis estimates, 170–71, 177–79, 238(n4). *See also* carbon dioxide; photosynthesis; water

oxygen catastrophe, 74

ozone (O$_3$), 72

Pacific Ocean: carbon-14 to carbon-12 ratio in, *156*, 156–57, 159; circulation, 148; climate records in, 213, 241(n15);

salinity, 143, 149; surface water age, 157–60

palynology, 204–5

paper chromatography, 85–87, 97–98, 234(ch6n2)

particle interceptor traps, 180

particles, atomic. *See* atomic particles

peat, 204–5

periodic table of elements, 6, 229. *See also specific elements and isotopes*

Peteet, Dorothy, 207–8, 240(n3), 241(n16)

petroleum, 56, 58. *See also* fossil fuels

PGA. *See* phosphoglyceric acid

pharmaceutical industry, 99

phosgene gas, 30

phosphates, 208, 240(n8). *See also* ribulose bisphosphate (RuBP)

phosphoglyceric acid (PGA), 88–89, *89*, 90–92, 93, *95*, 97. *See also* photosynthesis

phosphorescence, 38–39

photography (film), 38–39, 232(ch3n1)

photomultiplier tube (PMT), 104, *105*

photorespiration, 93

photosynthesis, 36; anoxygenic photosynthesis, 73–74; big bag experiments, 179–80, 188–92, 239(n15); biochemical process (CBB cycle), 88–96, *89*, *91*, *95*; Calvin team's work on carbon fixation process, vii, 28, 81, 83–93, *87*, 96–99, 168; carbon-14 assimilation and incubation length, 197; carbon-14 taken up, 54, 64; chlorophyll and, 84, 95–96; discovered by Ingenhousz, 77–79, 234(ch5n5); *Endeavor*'s ocean phytoplankton experiment (1991), 1–3; evolution of, 71–72, 74; history of scientific inquiry, before carbon-14, 76–80; importance of, 70, 72; Kamen's and Ruben's research, 19–20, 28, 80,

81–83, 223; light and darkness and, 83,
92, 95–96; microbial food web and,
199, 200; and ocean CO_2, 162, 172;
as ocean fertility indicator, 168, 170,
177–82, 223–24; ocean photosynthesis
studied using carbon-14, 172–78,
197, 223–24, 238(nn8, 10); overall
equation, 79; overall scheme, *75*;
oxygen measurement method
for estimating, in ocean, 170–71,
177–79; oxygen's role in, 80, 81;
phosphoglyceric acid (PGA), 88–89,
89, 90–92, 93; ribulose bisphosphate
(RuBP), 89–92, *91*; by zooxanthellae,
214. *See also* cyanobacteria; ocean
fertility; phytoplankton; plants; *and
specific scientists*
phytoplankton, 168–69, *169*; carbon-14
taken up by, in research study, 100;
challenges of studying, 167–68;
Endeavor's ocean phytoplankton
experiment (1991), 1–3; most ocean
photosynthesis performed by, 162;
nutrients, 155; ocean photosynthesis
studied using carbon-14, 172–78,
197, 223–24, 238(nn8, 10); oxygen
measurement used to estimate ocean
photosynthesis, 170–71, 177–79;
photosynthesis as indicator of ocean
fertility, 168, 170. *See also* ocean
fertility; photosynthesis; plankton
Pia, Secondo, *117*, 117–18. *See also* Shroud
of Turin
pigments, 71, 73, 84. *See also* chlorophyll
pipettor, 191
planetary vorticity, 146–47
plankton, 163, 168–70, *169*; Haeckel-
Hensen controversy, 179, 238–39(n13);
Steemann Nielsen's carbon-14 method
and, 175–76 (*see also* Steemann
Nielsen, Einer). *See also* foraminifera;
phytoplankton; zooplankton

Plankton Rate Processes in Oligotrophic
OceanS. *See* PRPOOS
plants, 75, 76–78, 83. *See also*
photosynthesis; pollen
platinum (Pt), 10–13, 229
PMT, 104, *105*
Poisson statistics, 110
pollen, 204. *See also* Younger Dryas
polonium-210, 49
Powell, Dick, 44
precursor/end-product experiments, 99
Priestley, Joseph, 77, 78
Prochlorococcus, 75
proteins, 36, 223
Protista, 84, 162, 169. *See also* algae;
foraminifera
protons, 6, *33*, 33–34, *34*
PRPOOS (Plankton Rate Processes in
Oligotrophic OceanS), viii, 181–82,
185–86, 192–93, 198–200. See also
Melville, RV
"pulse-chase" experiments, 67, 161
Purdie, Duncan, 199, 239(n4)

Quaternary Isotope Laboratory (Univ. of
Washington), 69
Quay, Paul, 156–57
quenching, in a scintillation counter,
105–7

Rabinowitch, Eugene, 79–80
radiation: assessing exposure, 40;
background radiation, 48–49,
56–58; dangers of, 39–40, 42–44,
48, 49, 232(ch3n4); electromagnetic
spectrum, *47*, 47; ionizing radiation,
47, 47–48, 49; units of measurement,
40. *See also* radioactivity
radioactivity: of all living things, 54;
autoradiography, 86–87; background
radiation, 48–49, 56–58; of carbon-14,
7; detecting and measuring, 40–41

radioactivity (*continued*)
  (*see also specific detectors*); discovery,
  7, 38–39; disintegrations per minute
  (dpm), 56; explained, 7, 40; half-life
  (generally), 7–8, 41 (*see also under*
  carbon-14); Kamen's exposure to,
  14, 22, 28, 37; particle types, 41–42,
  232(ch3n3); radioactive decay, 7–8, 41,
  48, 107 (*see also and specific elements
  and isotopes*); of scintillation counter
  standards, 102, 106. *See also* radiation;
  radioisotopes; *and specific substances*
radioisotopes: dating fossil coral cores
  by, 214, 241(n13); half-life, 7–8, 41;
  medical uses and demand, 17–18,
  20, 27, 49; produced by atmospheric
  nuclear tests, 160 (*see also* nuclear
  tests); radiometric dating using, 62
  (*see also* carbon-14 dating); wartime
  demand for, 13. *See also specific
  elements and radioisotopes, such as*
  carbon-14
radium (Ra), 39–40, 48, 155, 229
radon (Rn; gas), 40, 48, 229
rain rate, 180–81, 213
rat experiments, 19
reduction (defined), 73
redwood trees, 59, 68, 211
Renger, Ed, 191
research vessels, 1–3, 135–36, 140–41, 151,
  153–54, 174–75. See also *Melville*, RV;
  oceanographic tools and methods;
  oceanography
reservoir effect, 65, 129
Revelle, Roger, 28, 113
ribulose bisphosphate (RuBP), 89–92, *91*,
  93–94, *95*, 97
Riley, Gordon A., *171*, 238(n5); oxygen
  flux experiments and photosynthesis
  estimates, 171, 178–79, 193, 195,
  238(n9); trophic analyses, 171–72, 181,
  196

Rind, David, 207–8
RNA (ribonucleic acid), 36
Roentgen, William, 38
Ruben, Sam, *12*; collaboration with
  Kamen, 14; and the discovery of
  carbon-14, 6, 21, 22–24, 25; and
  Kamen's nitrogen-14 bombardment
  experiment, 31–33; and Lawrence's
  platinum–deuteron bombardment
  experiment, 12–13; life and death,
  12, 26, 29–30, 232(ch2n2); long-
  lived radioactive isotopes sought,
  20; photosynthetic carbon fixation
  research, 19–20, 80, 81–83
Rubisco (ribulose bisphosphate
  carboxylase/oxygenase), 92–93, 94, *95*
RuBP. *See* ribulose bisphosphate (RuBP)
Rutherford, Ernest, 7, 41, 227
Ryther, John, 179, 238(n3)

salinity. *See under* ocean
Sargasso Sea, 178, 179, 193–95, *194*
satellites, 198
Schou, Lise, 173
scintillation counter. *See* liquid
  scintillation counter
Scripps Institution of Oceanography,
  62–63, 98, 155, 181–82, 183, 193.
  See also Eppley, Dick; *Melville*, RV;
  PRPOOS; Suess, Hans
sediment varves, 69, 124, 215, 216
Segré, Emilio, 17, 21, 31, 227
Senebier, Jean, 78–79
sharks, 131
Shroud of Turin, 116–22, *117*, 235(n3)
Sieburth, John, 186
silica, 163
silicon (Si), 36, 229
slave ships, 138–39
soft-money science, ix–x
solar wind, 64, 65. *See also* cosmic rays
Soriano, Juan, 1–3

Southon, John, 113
speleothems, 69, 214–15
Spencer, Derek, 155
spinthariscope, 103, 234(ch7n2)
stalagmites and stalactites. *See* speleothems
statistics: Poisson statistics, 110; scintillation counters and carbon-14 measurement, 106, 107–8; standard deviation, 235(n2)
Steemann Nielsen, Einer, *173*, 239(n14); carbon-14 method for assessing ocean photosynthesis, 172–77, 196, 223–24, 238(n10); controversy over carbon-14 method, 177–79, 181, 193, 195, 196–97; sampling stations/locations, 174, *175*, 178, 193, 195
Steno, Nicholas, 124
Stepka, William, 85
Stommel, Henry (Hank), 238(n2); biological oceanography synthesis (with Riley, Bumpus), 178, 193; work on ocean circulation, 142, 147, 150, 151, 152, 154–55, 160
Stonehenge, 125
strontium (Sr), 49, 160, 229; strontium sulfate, 162
Stuiver, Minze, 68–69, *156*, 156–57, 241(n14)
submarines, 108–9
Suess, Hans, 62–63, 65, 68, 113, 126
Suess effect, 65–66, 113, 161
sugars, 19, 35, 81, 89–91, *91*, 93–94, 96, 164. *See also* photosynthesis; ribulose bisphosphate (RuBP)
Suigetsu, Lake, 215
sun: and ocean temperature and circulation, 142; ultraviolet light from, 72; variations in radiation from, 64–65, 68–69, 207, 225. *See also* cosmic rays; light
Sverdrup, Harald, 241(n9)

Swallow, John, 151
Swallow floats, 150, 151–52
Swift, Elijah, 187, 188
Synge, R. L. M., 85, 234(ch6n2)
Szilard, Leo, 14

tandem electrostatic accelerator, 112. *See also* accelerator mass spectrometry
tannins, 165
technetium (Tc), 17, 41, 229, 231(n2)
thermocline, 141
thermohaline circulation (TCH), 142–43
Thompson, Benjamin (Count Rumford), 139–40
thorium (Th), 62, 229, 231(n2), 241(n12)
three-dog night, 205, 240(n4)
time-series stations, 179, 238(n2)
trade winds, 145
Transient Tracers in the Ocean (TTO), 237(n10)
tree rings: and carbon-14 dating, 59, 63–64, *66*, 68, 69, 123; and climate chronologies, 211–12
trophic structure (defined), 170
Trumbore, Susan, 113
Tyler, Tom (Capt.), 2–3

United States Navy, 108–9
University of California (UC) Irvine. *See* Keck Carbon Cycle Accelerator Mass Spectrometry Laboratory
University of California, San Diego (UCSD), 28, 62. *See also* La Jolla Radiocarbon Laboratory
uranium (U), 38–39, 62, 69, 229, 231(n2), 241(n12)
Urey, Harold, 20, 62, 227

Vaccaro, Ralph, 179
valence (defined), 34
van Helmont, Jan Baptist, 76, 79
variolation, 77, 234(ch5n4)

Vema (research vessel), 153–54, 237(n4)
von Hevesy, George, 173, 174, 227
vorticity, 146–47

Warburg, Otto, 96–97, 227
water (H$_2$O), 72; anoxic waters, 215, 216;
    carbon's solubility in, 35; groundwater
    and speleothems, 214–15; used in
    photosynthesis, 72. *See also* lakes;
    ocean; ocean circulation; ocean fertility
water samplers, large-volume, 150
Wayne, John, 42, *43*, 44
weather ships, 179
Weiss, Ray, 237(n8)
whales, 170
Wildman, Sam, 92
Williams, Peter J. leB., 199, 239(n4)
Wilson, Charles, 9. *See also* cloud chamber
Wilson, Robert, 21, 30–31
winds, and ocean circulation, 142–45, 194
Winkler, Ludwig Wilhelm, 170, 177;
    oxygen measurement technique,
    170–71, 177–79
wood, carbon-14 dating of, 59, *60*. *See
    also* tree rings

Woods Hole Oceanographic Institution,
    194; AMS at, 235(ch7n7); focus on
    Northwest Atlantic, 193–94; and
    GEOSECS, 155 (*see also* GEOSECS);
    ocean circulation research, 151;
    oceanographers' backgrounds,
    238(n2); Riley and, 171, 179. *See also*
    Broecker, Wally; Riley, Gordon A.;
    Stommel, Henry
World Ocean Circulation Experiment
    (WOCE), 237(n10)
World War II, 26–28, 29–30. *See also*
    Manhattan Project
Worthington, Val, 238(n2)

X-rays, 38–39, *47*, 48, 232(ch3n2)

Younger Dryas (climate period), 204–7,
    206, 209–10, 215, 217–22, *218, 220,*
    240(n3), 241(n17). *See also* climate and
    climate change

zooplankton, 100, *169*, 169–70, 171, 175.
    See also plankton
zoozanthellae, 214